Roberto Quesquén Fernández
José Rivasplata Cruz

Introduction to Oceanography

Roberto Quesquén Fernández
José Rivasplata Cruz

Introduction to Oceanography

With emphasis on physical oceanography

ScienciaScripts

Imprint

Any brand names and product names mentioned in this book are subject to trademark, brand or patent protection and are trademarks or registered trademarks of their respective holders. The use of brand names, product names, common names, trade names, product descriptions etc. even without a particular marking in this work is in no way to be construed to mean that such names may be regarded as unrestricted in respect of trademark and brand protection legislation and could thus be used by anyone.

Cover image: www.ingimage.com

This book is a translation from the original published under ISBN 978-620-2-15205-1.

Publisher:
Sciencia Scripts
is a trademark of
Dodo Books Indian Ocean Ltd. and OmniScriptum S.R.L publishing group

120 High Road, East Finchley, London, N2 9ED, United Kingdom
Str. Armeneasca 28/1, office 1, Chisinau MD-2012, Republic of Moldova, Europe
Printed at: see last page
ISBN: 978-620-7-66875-5

Contents

CHAPTER I

INTRODUCTION

The oceans have always aroused man's curiosity and interest in understanding them. Throughout history, the ocëano has had a strong impact on man's life. Although this principle is widely accepted, it is often little understood. Moreover, its study is complex because this system, the ocean, is closely interrelated with other systems such as the atmosphere and the biosphere. For this reason, it is necessary to call upon other sciences such as meteorology, ecology, geo-economics, physics, chemistry, etc.

The knowledge achieved is allowing the formulation of some mathematical models that predict quite accurately a few months in advance the occurrence of phenomena and/or alteration of the main parameters of the ocean.

The Earth is a system composed of subsystems that are interacting with each other in a complex way. These are the geosphere, the hydrosphere and the atmosphere, each of which has given rise to a science, namely geology, oceanography and meteorology, all known as the earth sciences. These sciences approached in this way have been very successful in explaining and understanding our planet Earth.

These three components constitute moving fluids whose main difference is their viscosity. The geosphere has the highest viscosity and is therefore much slower moving.

The ocean and the atmosphere are the subsystems with the most interacting elements that shape the dynamics of the oceans and the phenomena that directly affect them.

Since it is ultimately in man's interest to know the oceans in order to use the resources they contain in a suitable way, managers of these resources cannot ignore the laws of nature, as they form the basis for any environmental management.

The present work, an elaboration of the text - Introduction to Oceanography, is the first part of a larger text project. The first four chapters of this text include basic and complementary aspects necessary for a clear understanding of the later chapters. Thus, after the present introduction, chapter 2 will present the basic aspects of the science of physics such as the dimensions and units most commonly used in oceanography, the physical laws that are fundamental for the understanding of the main oceanographic professions and some properties of this science such as gravity, pressure and mass fields. Chapter 3 covers the different elements of the atmosphere and the oceans that interact with each other and that are important for characterising the dynamics of water masses. Chapter 4 deals with the equilibrium heat flux to the Earth and general aspects of the heat distribution process and its impact on the oceans.

In Chapter 5 we will begin to deal with the ocean itself, so the first topic is ocean variables, i.e. temperature, salinity and pressure. We will define them, discuss their origins, their variability and distribution in the oceans, as well as the units used. Useful methods of representing these variables will be presented.

Chapter 6 deals with the physical properties of the oceans. Here the main properties such as density, as well as the thermal properties of the sea such as heat capacity and others will be discussed. We will also look at their distribution, their variability in space and time in the oceans.

Chapter 7 studies the surface layer of the oceans, the processes that occur in it, the influence of winds on this layer and the interaction that occurs there. The factors that delimit them and the factors that generate them. Finally, in chapter 8, perhaps the most extensive and the most studied chapter on the oceans because of the impact it has on all oceanographic processes, how the circulation is generated and what are the major circulations in the oceans.

This text will aim to present these topics in a clear and understandable way in order to serve as a reference for students of fisheries and related subjects. As far as possible, it will avoid the use of mathematical and physical explanations that involve complex equations or require a deep knowledge of mathematics, but without losing the corresponding scientific rigour.

CHAPTER II

BASIC PHYSICS FOR OCEANOGRAPHY

The understanding of the oceans, their properties, the processes occurring in ёl can often be explained and understood by relying on basic sciences such as mathematics and physics. Various principles and laws of these sciences are frequently used in oceanography. In this chapter we will recall the most frequently used ones in this field. First it is necessary to define both units and dimensions.

The word dimension usually refers to the size of things, but in the science of physics it represents the fundamental category by which the characteristics or properties, processes or physical bodies are described. In this case no numerical magnitude is involved. Now, in mechanics and hydrodynamics, these fundamental dimensions are mass, length and time which are designated with the symbols M, L and T. However, in their characterization they are expressed with units accompanied by numerical quantities. There are two types of units, fundamental and derivative.

Fundamental units. The generally accepted units of mass, length and time are the gram, centimetre and second, thus grouped together they are known as the c.g.s. system. In oceanography it is not always practical to use these units, so for measuring depths it is more convenient to use the metre as a unit, for masses it is more convenient to use tons, for time the second is used (m.t.s. system). For tërmic processes, degrees Celsius (°C) is used as a unit. When measuring horizontal distances it is more convenient to measure in nautical miles or kilometres. For that reason in oceanography it is necessary to indicate the units in which you are measuring.

Derived units. In mechanics, derived units are expressed in the three dimensions M, L and T and by the values of the units used. Thus, velocity has the dimension length divided by time, which is written as LT^{1} and is expressed in centimetres per second or metres per second. Velocity can in fact be expressed in many other units such as miles per hour or miles per day, but the dimensions are the same.

Term	Dimension	Unit in c.g.s. system	Unit in m.t.s. system
Fundamental f-unit Mass Length Time *Derived unit* Speed Acceleration Angular velocity Momentum Momentum Force Impulse Work Kinetic energy Activity (energy) Density Specific volume Pressure Gravitational potential Dynamic Viscosity Kinematic Viscosity Diffusion	M L T LT-1 LT-2 T-1 MLT-1 MLT-2 MLT-1 MLT-1 ML2T-2 ML2T-2 ML2T-2 ML2T-3 ML-3 M-1L3 ML-1T-2 L2T-2 ML-1T-1 L2T-1 L2T-1 L2T-1 L2T-1	g cm s cm/s cm/s2 1/s g cm/s g cm/s2 = 1 dyne g cm/s g cm2/s-2 = 1 erg g cm2/s2 = 1 erg g cm2/s3 = 1 erg/s g/cm3 cm /gᵌ g/cm/s2 = dyne/cm2 cm2 cm2/s2 g/cm/s cm2/s cm2/s cm2/s	metric ton = 106 g metre = 102 cm s m/s = 100 cm/s m/s2 = 100 cm/s2 1/s ton m/s = 108 g cm/s ton m/s2 = 108 dynes ton m/s = 108 g cm/s ton m2/s2 = 1 kilojoule ton m2/s2 = 1 kilojoule ton m2/sᵌ = 1 kilowatt ton/mᵌ = g/cm3 m3/ton = cm3/g ton/m/s =1 centibar m2/s2 = 1 dynamic decimetre ton/m/s = 104 g/cm/s m2/s = 104 cm2/s m2/s = 104 cm2/s

The table shows the dimensions of some terms that are frequently used in oceanography. Certain terms have the same dimensions, although the concepts differ.

In physical equations, all tërms must have the same dimensions with equal exponents. However, in some relations it is often not quite exact, for example, when saying that the acceleration of a body is equal to the sum of the forces acting on this body, when comparing their dimensions we have that the acceleration is LT-, while the force is MLT^2 . The correct way to say is that the acceleration of a body is equal to the sum of forces per unit mass acting on that body. An example of correct ways of expressing the physical relations is the pressure exerted by a column of water of constant density p, and height h at a place where the acceleration of gravity is g.

$$p = \rho g h$$

$$ML^{-1}T^{-2} = ML^{-3} \times LT^{-2} \times L$$
$$ML^{-1}T^{-2} = ML^{-1}T^{-2}$$

Some of the constants appearing in the physical equations have dimensions and their numerical values depend on the particular units that have been assigned to the fundamental dimensions while other constants are dimensionless and therefore independent of the system of units. Density has dimensions ML^{-3} , but the density of pure water at 4° has the numëric value of 1 only if the units of mass and length are selected in a special way (grams and centimetres or metric tons and metres). On the other hand, the specific gravity, which is the density of a body relative to the density of pure water at 4° is dimensionless (ML^{-3} / ML^{-3}) and is expressed by the same number, depending on the system of units used.

2.1. PHYSICAL LAWS USED IN OCEANOGRAPHY

They are:

1. Mass conservation.
2. Energy conservation.
3. Newton's First Law of Motion, which tells us that if there is no resultant force acting on a body there will be no change in the body's motion.
4. Newton's Second Law of Motion, in which the change in the change of motion is directly proportional to the resultant force acting on it and is in the direction of that force.
5. Newton's Third Law of Motion, which states that for any force acting on one body there is an equal and opposite force acting on another body.
6. Conservation of angular momentum.
7. Newton's law of gravitation.

Strictly speaking the first two laws are related, but for our case where we are not involved with the Mass ^ Energy conversion, it is convenient to keep them separate. The conservation of mass is fundamental, but in oceanography we mostly use the so-called continuity equation, which expresses the conservation of mass/unit volume (density) or volume. Laws 3, 4 and 5 are aspects of linear momentum conservation.

The two types of energy whose conservation is important in oceanography are heat and mechanics. Heat conservation, or heat balance, is most important when discussing temperature distribution as a property of seawater. On the other hand, the conservation of mechanical energy is considered when dealing with waves, while the conversion of mechanical energy to heat is considered as a loss process which will not be discussed here.

The ocean is dynamic, i.e. it is in constant motion. In some cases, motion occurs under a system of forces in equilibrium such that no resultant force acts, this is the case covered by Newton's First Law of Motion. In other cases there is a resultant force and acceleration occurs, the relationships between force and acceleration are determined by Newton's Second

Law. Angular momentum occurs in vorticity. The Law of Gravitation, as it stands, is applied primarily in ocean tidal dynamics, although it is also important in determining the distribution of hydrostatic pressure and the cause of motion when density change occurs.

An example of the application of these laws is the equation describing motion in the ocean which is derived from Newton's Second Law, which expresses conservation of momentum (i.e. the product of mass times velocity) in the following form:

$$\mathbf{F} = m\,\mathbf{a}$$

Bold letters indicate vectors and italic letters indicate scalar values. In fluids this equation is expressed in tërms of force per unit mass:

$$\mathbf{F'} = \frac{\mathbf{F}}{m}$$

from which it follows that:

$$\mathbf{F'} = \frac{d\mathbf{v}}{dt}$$

here $\mathbf{v} = (u,v,w)$ *is the* velocity expressed in its components along the *x, y, and z* axes (*x* eastward, *y* northward and *z* downward). Recall that the acceleration (**a**) is equal to **dv/dt**.

If there is more than one force, Newton's Second Law applies to the sum of all the forces involved. The law is valid in an absolute coordinate system, i.e., a stationary system or one moving at constant velocity. In oceanography, coordinate systems are generally defined with their origin somewhere on the surface of the Earth, so they are neither stationary nor moving with constant velocity, but rather rotate along with the Earth. To solve this problem, an apparent force is included in Newton's Second Law to account for the effects of rotation. By adding up all the forces acting on the ocean, Newton's Second Law takes the form:

particle acceleration = -pressure gradient + Coriolis force + tidal force per unit mass + friction + gravity

The tidal force is only considered in tidal problems so it is not necessary to include it in the general ocean circulation. On the other hand, the force of gravity does not exert any horizontal force and therefore cannot produce any horizontal acceleration and is only important when considering motions that include vertical motion such as convection, waves, etc.

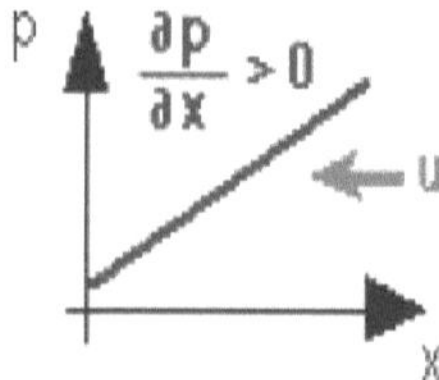

Figure 2.1

From the above formula, the pressure gradient is negative ^why? Looking at the attached figure (Figure 2.1), the pressure *p* increases as the distance *x* increases (to the right), the pressure gradient is positive, but the acceleration is from the high pressure to the low pressure so the current *u* flows down the pressure gradient (to the left).

The Coriolis force is an apparent force and exists only for an observer in a rotating reference system. In oceanography, currents are always expressed with respect to the bottom of the ocean (which rotates with the Earth) and can therefore only be correctly analysed if the Coriolis force is taken into account in the summation of forces.

2.2. GRAVITY, PRESSURE AND MASS FIELDS

Usually the ocëano is characterised by its geometrical dimensions. But in problems of statics and dynamics involving the action of forces this is not sufficient. For this, it is convenient to take as a reference surface levels which at any of its points is normal to the force of gravity (Sverdrup, 1942) and does not always coincide with the surfaces of equal geometrical depth. Under these conditions, when only the force of gravity acts on a mass, it moves along a surface level without expenditure of work and that the amount of work expended or gained in moving a unit mass from one surface to another is independent of the route taken. The amount of work, W, required to move a unit of mass a distance, h is:

$$W = g\,h$$

where g is the acceleration of gravity. The work per unit mass has dimensions L^2 $T2$ so the numerical value depends only on the units used for length and time. When length is measured in metres and time in seconds, the unit of work per unit mass is called the *dynamic decimetre*. If we consider the sea surface to be a surface level. The work required or gained in moving a unit of mass from sea level to a point above or below is called *gravitational potential* and in the m.t.s. system the practical unit of gravitational potential is the *dynamical metre*, (although the most commonly used is the dynamical decimetre) and its symbol is -*DII*. With respect to the sea the vertical axis is taken as positive downwards. Therefore, the geopotential of a level surface at geomëtric depth z, in dynamic metres is:

$$D = \frac{1}{10} \int_0^z g\,dz$$

On the other hand, the geometric depth in metres of a given level surface is:

$$z = 10 \int_0^D \frac{1}{g}\, dD$$

The gravity field. Gravity is the result of two forces, the attraction of the earth and the centrifugal force due to the rotation of the earth. Furthermore, it is not necessary to take into account the small irregular variations of gravity, it is sufficient to take the normal value, in metres per second, which at sea level can be represented as a function of latitude (φ). The normal value at the poles is 9.83205 and at the equator is 9.78027. Furthermore, according to Bjerknes (1941) gravity increases with depth. Consequently, the geopotential, in dynamic metres, for a given depth, z:

$$D = \frac{g_0}{10} z + 0,1101 \times 10^{-6} z^2$$

To determine the depth corresponding to a given value of D:

$$z = \frac{10}{g_0} D - 0,1168 \times 10^{-6} D^2$$

In the first approximation, $D = 0.98z$ and $z = 1.02D$. This means that the value representing the depth in metres deviates by only about 2% from the value representing the geopotential in dynamic metres. It should be kept in mind that a dynamic metre is a measure of work per unit mass (LT^2) and not a measure of length (L T).$^{-12}$

The gravity field can be described by a set of equipotential surfaces corresponding to standard gravity potential intervals. There are equal distances if the geopotential is used as the vertical coordinate, but if the geometric depth is used, the distance between the equipotential surfaces varies. If the field is represented by equipotential surfaces at intervals of a dynamic decimeter, it follows from the definition of these surfaces that the value питёпсо of the acceleration of gravity is the reciprocal of the geomëtric thickness, given in metres of the unit layer.

The depression field. The pressure distribution in the sea can be determined by the static equilibrium equation:

$$dp = k\rho_{s,t,p}gdz$$

Aqrn, k is a numerical factor depending on the unit used and ps,t,p is the density of water. So far as aqw, it is sufficient to emphasise that, as far as conditions permit, the equation for all practical purposes is exact.

Entering the geopotential, expressed in dynamic metres as the vertical coordinate, gives $10dD$ = gdz. When the pressure is measured in *decibars* (defined by 1 bar = 10^6 dynes per cm^2), the factor k is then equal to $1/10$ and the equation reduces to:

$$dp = \rho_{s,t,p}dD \quad \text{o} \quad dD = \alpha_{s,t,p}dp$$

where $\alpha_{s,t,p}$ is the specific volume.

Since density and specific volume differ little from unity, a pressure difference in decibars has almost the same numerical value as the geopotential difference in dynamic metres or the difference in geomëtric depth in metres:

$$p_1 - p_2 = D_1 - D_2 = z_1 - z_2$$

A system of isobaric surfaces allows a good description of the pressure field. Indeed, using the geopotential as the vertical coordinate, the pressure distribution can be presented by a series of maps with isobars on standard level surfaces or by a series of maps of the geopotential topography of standard isobaric surfaces. In meteorology, the first form is generally used for weather maps, in which the pressure distribution at sea level is represented by isobars. In oceanography, on the other hand, it has been found practical to represent geopotential topography on isobaric surfaces.

The pressure gradient is defined by:

$$G = -\frac{dp}{dn}$$

where n (of dimension L) is in the direction normal to the isobaric surface. The pressure gradient has the character of a *force per unit volume*, since the pressure has the dimensions of a force per unit area (ML T^{-1-2} = MLT^{-2} x L^{-2}) and the dimension of a pressure gradient is ML T^{-2-2} = MLT^{-2} x L^{-3} . Multiplying the pressure gradient by the specific volume, M^1 L ₃, gives a force per unit mass of dimension LT^{-2} .

The pressure gradient has two main components: the vertical component, with direction normal to the level surfaces, and the horizontal component, with direction parallel to the level surfaces. When static equilibrium exists, the vertical component, expressed as force per unit mass, is balanced by the acceleration of gravity (obtained mathematically by means of the equation of hydrostatic equilibrium). In a system at rest the horizontal component has no one to balance with and therefore there is a horizontal pressure gradient, which means that the system *is not at rest* or *cannot remain at rest*. For this reason, the horizontal pressure gradient, although extremely small, is important in establishing the state of motion, whereas the vertical pressure gradient is negligible in this respect.

It is evident that there is no movement due to the pressure distribution if the isobaric surfaces coincide with the level surface. In this state of perfect hydrostatic equilibrium, the horizontal pressure gradient disappears and would only be present if the atmospheric pressure were constant acting on the sea surface, if the sea surface coincides with the ideal sea level and if the density of the water depends only on the pressure.

None of these conditions is fulfilled. The isobaric surfaces are generally inclined relative to the level surface, and the horizontal pressure gradient is present, forming an internal force field.

This force field can also be defined by considering the inclinations of isobaric surfaces instead of horizontal pressure gradients. By definition, the pressure gradient along an isobaric surface is zero, but, if this surface does not coincide with a level surface, a component of the acceleration of gravity acts along the isobaric surface and will tend to set the water in motion, or must be balanced by other forces if a steady state of motion is reached. The internal force field can therefore also be represented by the component of the acceleration of gravity along the isobaric surfaces.

For a complete description of the force associated with the pressure distribution, it is necessary to know the *absolute* isobars on level surfaces or the *absolute* geopotential contour lines of isobaric surfaces. It is not possible to satisfy these requirements. One reason is that measurements of the geopotential distances of isobaric surfaces must be made from the actual sea surface, the topography of which is unknown. In practice, what can be done is to determine the pressure distribution that would be present if it depends only on the mass distribution in the sea. This part of the total pressure field is called the *relative pressure field*, but it must be remembered that the *total pressure field* is composed of the relative field and a field due to external forces such as atmospheric pressure and wind.

By means of density observations at different depths, the relative pressure field can be derived and can be represented by the topography of the *relative* isobaric surface with some of the arbitrarily selected isobaric surfaces or not. The *relative* force field can be derived from the inclination of the isobaric surface with respect to the selected reference surface, but in order to find the absolute pressure field and the corresponding absolute force field, it is necessary to determine the absolute shape of an isobaric surface.

These considerations have been taken into account in detail because it is essential to be fully aware of the difference between the absolute pressure field and the relative pressure field and to know what types of data are needed to determine each of these fields.

The mass field. The mass field in the océano is generally described by means of the specific volume, expressed by:

$$\alpha_{s,t,p} = \alpha_{35,0,p} + \delta$$

the specific volume field can be considered as composed of two fields, the field of $\alpha_{35,0,p}$ and the field of δ.

The first field is of a single character. The surface of $\alpha_{35,0,p}$ corresponding to the standard ocëano coincides with the isobaric surface, the deviations of this from the level surface are so small that for practical purposes the surfaces $\alpha_{35,0,p}$ can be considered as coincident with the level surfaces or with surfaces of equal geomëtric depth. The field of $\alpha_{35,0,p}$ is therefore fully described by means of tables given for $\alpha_{35,0,p}$ as a function of pressure obtaining the average relationship between pressure, geopotential depth and geomëtric depth. Since this field is considered constant, the mass field is described by means of the anomaHa of the specific volume, δ.

The mass field is represented by means of the topography of the anomaHa surfaces or by means of horizontal maps or vertical sections in which the curves of δ can be placed as a constant. The latter method is the most common. It should always be kept in mind, however, that the *in situ* specific volume is equal to the sum of the standard specific volume, $\alpha_{35,0,p}$, at the *in situ* pressure and the anomaHa, δ.

The relative pressure field. It is impossible to determine the relative sea pressure field by direct observations, because an error of only 0.1 m in depth, any manometer introducing errors greater than the existing horizontal differences.

However, if the mass field is known, the internal pressure field can be determined from the equation of state equilibrium in one of the forms:

$$dp = \rho dD \quad \text{o} \quad dD = \alpha dp$$

In oceanography the latter form is more practical, although both are applicable. The integration of the latter form gives:

$$D_1 - D_2 = \int_{p_1}^{p_2} \alpha_{s,\vartheta,p}\, dp$$

since:
$$\alpha_{s,\vartheta,p} = \alpha_{35,0,p} + \delta$$

then you can write

$$\left(D_1 - D_2\right)_s + \Delta D = \int_{p_1}^{p_2} \alpha_{35,0,p}\, dp + \int_{p_1}^{p_2} \delta dp$$

From the equation it follows that the relative pressure field is composed of two fields: the standard field and the anomaly field. The standard field is determined once and for all cases, if the salinity of the sea is constant at 35^ and the temperature is constant at 0°C. The standard geopotential distance decreases with increasing pressure, because the specific volume decreases (density increases) with pressure, according to which the standard geopotential distance between the isobaric surfaces of 0 and 100 decibars is 97.242 dynamic metres. Moreover, they are independent of latitude, unlike the geomëtric distance which does depend on g varies. On the other hand, the above equation shows that:

$$\left(D_1 - D_2\right)_s = \int_{p_1}^{p_2} \alpha_{35,0,p}\, dp$$

y is called the standard geopotential distance between the isobaric surfaces of *p1* and *p2* and where:

$$\Delta D = \int_{p_1}^{p_2} \delta dp$$

This expression is known as the geopotential distance *anomophany* between the isobaric surface p_1 and *p2* or, in short, the *geopotential aiiomalia*. In order to evaluate this equation, it is necessary to know the anomophysia, 5, as a function of absolute pressure. The anomafia is calculated from observations of temperature and salinity, although oceanographic observations give information about temperature and salinity at known geomëtric depth below the real sea surface, and not at known pressure. These difficulties can be overcome by artificial substitution, because at any given depth the absolute pressure value expressed in decibars is almost the same as the depth value, as is evident from the following corresponding values:

Standard sea pressure (decibars   )1000 2000 3000 4000 5000 6000
Approx. geomëtric depth (metres)   990 1975 2956 3933 4906 5875

Thus, the values of the geomëtric depth deviate only 1 or 2% from the value of the standard pressure at that depth. This relationship is not accidental, although it is based on the use of a practical unit of pressure, the decibar.

It follows that the temperature at 1000 decibar pressure is almost equal to the temperature at the geomëtric depth of 990m. The vertical temperature gradient in the ocean is small, especially at great depths, and therefore no errors of consideration are introduced if, instead of using the temperature at 990 m when calculating 5, the temperature at 1000 m and so on can be used. The *difference* between the anomorphies of neighbouring stations will be less affected by this procedure, because in a given area the vertical temperature gradient is similar. The error introduced will be very similar at both stations and the difference will be a negligible error. In practice, *the numbers representing the geometric depth in metres* are considered as *the representation of the absolute pressure in decibars*. In this way it is possible to calculate, by means of tables, the anomorphism of the specific volume at the given pressure. To obtain the geopotential anomophysics of the isobaric layer in question, expressed in dynamic metres, is achieved by multiplying the average anomophysics of the specific volume between two pressures by the difference in pressure in decibars (which is considered to be equal to

the difference in depth in metres). By adding these geopotential anomalies, the corresponding anomaly between any pair of pressures can be found.

Since 5 decreases with depth, 51-62 is positive and the slope of one isobaric surface relative to another is of opposite sign to the slope of the surfaces 5. This rule allows a quick estimation of the relative slopes of the isobaric surface in a section in which the mass field has been represented by curves 5.

Isobaric surface profiles based on data from a series of stations in a section must obviously agree with the slope of the curve, as would be shown in a section and based on the same data, but this obvious rule often receives little or no attention.

Characteristics of the total pressure field From the above it is evident that, in the absence of a relative pressure field, the isostatric and isobaric surfaces must coincide. Therefore, if for some reason an isobaric surface, such as the free surface, deviates from a level surface, then all isobaric and isostatic surfaces must deviate in a similar manner. It is assumed that an isobaric surface under agitated conditions falls at a distance Δ' h cm below the position under non-agitated conditions. Then, any isobaric surface along the same vertical is also displaced from its position Δ h without agitation. The distance Δ h is positive downwards because the positive *z-axis* points downwards. The pressure at a given depth under no-stirring conditions is called p_0. The pressure under agitated conditions is $pi = p_o - \Delta$ p, where Δ p = gp Δ h and where it is considered that there is displacement due to the deficit or excess mass in the water column under consideration.

This consideration is equally valid if a relative pressure field exists. The absolute pressure distribution can always be determined from the equation:

$$p_i = p_0 - g\rho\Delta h$$

and therefore it would be known whether Дh, vertical displacement of the isobaric surface by excess or deficit of mass in the column under consideration, can be determined. An additional horizontal force field is present when this vertical displacement varies from one location to another, in which case the absolute isobaric surface has an inclination relative to the isobaric surface of the relative field. The additional field can be called the *tilt* field and this analysis thus leads to the result that the total pressure field is composed of the internal field and the *tilt* field. This distribution is very useful when discussing the character of the currents.

CHAPTER III

THE ATMOSPHERE AND ITS INTERACTION WITH THE OCEANS

The planet Earth is a complex system whose different elements interact with each other. Over the last 30 years, important advances have been made in the understanding of the interaction between the ocean and the atmosphere. The effects and causes of these interactions are complex and it is difficult to determine where the cause and effect occur. In the contact zone between the two, there is an exchange of suspended chemicals, energy (mainly heat) and momentum (winds). The circulation of the ocean depends mainly on two atmospheric factors: wind and seawater heating.

The ocean, which can store heat much better than the atmosphere or the land, absorbs more heat per unit area in the equatorial zone than at the poles. This heat will be transferred to the cooler areas of the ocean by convection or water movement. The heat storage capacity of the ocean is very important in modifying and influencing the continental climate.

The wind blowing over the ocean generates waves, mixes the surface water and extracts water vapour from the sea surface and eventually transferred to the land by precipitation. This cycle, called the hydrological cycle, is completed when the water returns to the ocean. Near the coast, wind can cause surface water to move offshore and be replaced by deeper, cooler, nutrient-rich water. This process is known as upwelling and can produce rich fishing grounds, because it transports nutrients from the bottom to the surface waters.

Brief description of the atmosphere and its general circulation

The atmosphere is a gaseous layer that surrounds the planet Earth. 75% of the atmosphere is found in the first 11 km of altitude from the planetary surface and 99% is concentrated in the first 30 km. Compared to the radius of the Earth (6400 km), the thickness of the atmosphere is very small, and the atmospheric thickness is very small.

Its chemical composition is influenced by the activity of the biosphere such as photosynthesis and respiration. It is also important in biogeochemical cycles, in the control of the climate and the environment in which we live, and it contains two of the three elements essential for life (nitrogen and carbon) in addition to oxygen. Man, with his own activity in today's society, is modifying the composition of the atmosphere, such as the increase of carbon dioxide or methane (causing the greenhouse effect), nitrogen oxide (causing acid rain) or the emission of fluorocarbon (responsible for the thinning of the ozone layer).

For ease of study, the atmosphere has been divided into layers: troposphere, tropopause, stratopause, mesopause and thermopause, exosphere. The **troposphere**, which extends up to an upper hemisphere called the **tropopause**, is on average about 12 kilometres (9 km at the poles and 18 km at the equator). Here, vertical and horizontal movements of air masses (winds) occur and there is a relative abundance of water, due to its proximity to the hydrosphere. In this zone are clouds and climatic phenomena such as rainfall, winds, temperature changes, etc. In the troposphere the temperature decreases with altitude, reaching -70°C in the upper hemisphere. The **stratosphere** starts from the tropopause and goes up to an altitude of 50 km. Here the temperature changes its tendency and rises to **~0°C**. There is almost no vertical air movement, but horizontal winds often reach 200 km/h, so any substance in this layer spreads rapidly around the globe, as is the case with ozone-destroying CFCs. Ozone is located between 30 and 50 km and absorbs harmful short-wave radiation.

The **mesosphere** extends from 50 km to 90 km. It is characterised by negligible concentrations of ozone and water vapour, a decrease in temperature, reaching -80 to -90°C at an altitude of 80 km, (lower than the previous two layers). As the earth's altitude increases, the atmosphere begins to be enriched with lighter gases. At this altitude the residual gases begin to stratify according to their molecular mass, due to a separation by the effect of gravity.

The **ionosphere** or **thermosphere** reaches up to 400 km in altitude. Its density is very low because the air is rarefied. AШ is where the northern lights occur and where certain frequencies of radio waves are reflected back to earth, but their functioning has little effect on living beings. Again, the tërmic level increases with altitude, especially between 120 and 150 km *(300° C),* and from 150 km onwards the increase is more gentle. Above 80 km, ultraviolet radiation, X-rays and electron rain from the Sun ionise several layers of the atmosphere, making them electrically conductive.

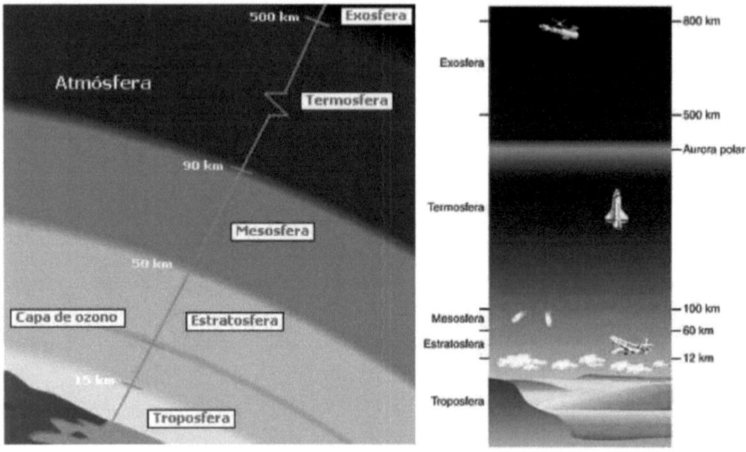

FIGURE 3.01. Source: http://encolombia.com/medioambiente/ Componentes_Medio.htm
FIGURE 3.02. Source: www.ieslosremedios.org/ ~elena/ websociales/index.html

The **exosphere** is the transition zone between outer space and the gaseous mass surrounding the planet. It is located between 900 and 9,600 km above sea level. It is made up of plasma. In it, the ionisation of the molecules makes the attraction of the Earth's magnetic field stronger than that of the gravitational field (which is why it is also called the magnetosphere). Therefore, the molecules of the lighter gases escape into interplanetary space without the Earth's gravitational force being sufficient to hold them back.

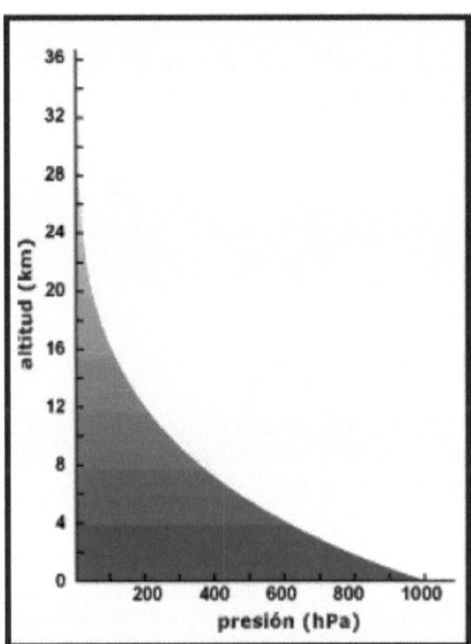

FIGURE 3.03. Source: http://www.atmosfera.cl/

The **atmospheric pressure** at a point corresponds to the force (weight) that the atmospheric column above that location exerts per unit area, due to the gravitational pull of the Earth. The unit used is the hectopascal (hPa) or millibar (mb). The average atmospheric pressure at sea level is slightly higher than 1000 hPa, i.e. about 10 tonnes per square metre (1 kg per cm^2). As the atmosphere is compressible, the effect of gravitational force causes its density to decrease with height, which in turn explains why the decrease in pressure with height is not linear, as shown in the figure below.

FIGURA 3.04.

The **general circulation of the atmosphere** has its origin in the energy radiated by the sun. The shape of our planet is spherical, so the sun's rays reach the surface in latitudinally

15

differentiated directions, giving rise to differentiated heating. Thus, when the sun's rays fall perpendicular to the surface in the equatorial zone, this is heated in greater proportion than in the polar zones where the sun's rays fall obliquely, covering a larger area for the same amount of heat.

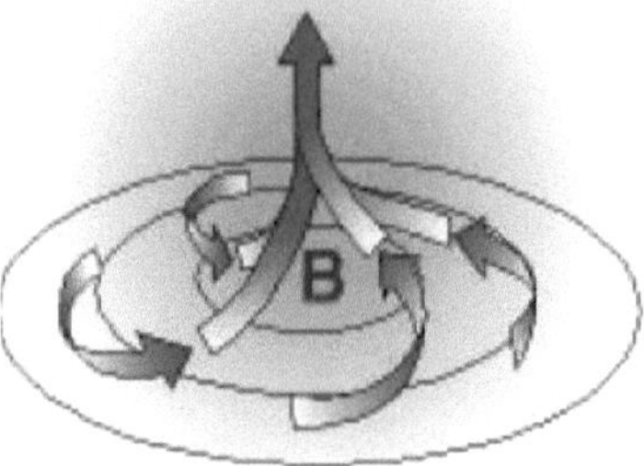

FIGURE 3.05. Source:
http://www.inamhi.gov.ec/educativa/ dictionary/letterC.htm

For practical purposes, the sun's rays do not transfer any heat as they pass through the atmosphere. When they reach the earth's surface it is heated, in the equatorial zone to a greater extent than at other latitudes of the planet. This hot surface radiates the heat to the surrounding atmosphere, warming it. Due to the property of gases, as the temperature increases, they expand and their density decreases, which causes them to rise, as shown in the attached figure. Since solar radiation is permanent, a continuous vertical upward flow of warm air masses is generated until it reaches the upper part of the troposphere. At this height the air masses diverge in a southward and northward direction (see figure below). If the Earth did not have a rotational motion on itself, then this airflow would reach the poles themselves. The rotational motion of the earth gives rise to the Coriolis effect, which tends to deflect all moving objects in the earth system. The air masses undergo a deflection in their long path breaking this circulation into three cells, known as, from the equator to the poles, Hadley Cells, Ferrer Cell and Polar Cell.

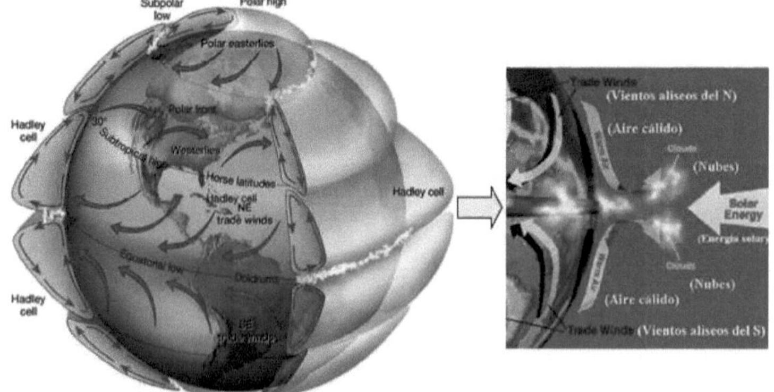

FIGURE 3.06. Source: http://www.ideam.gov.co/files/atlas/circulacion%20general%20en%20colombia.htm and www.dhn.mil.ve/noticia/noticia5.html

Winds then flow with a direction from the high atmospheric pressure zone to the low pressure zone. The direction is affected by the rotational motion of the earth and its speed by the friction with the surface of both continents and oceans.

The areas where rising air masses occur are known as low pressure zones and are characterised by their climatic instability (see figure 3.05). The equatorial strip where this vertical rise of air occurs is known as the Intertropical Convergence Zone (ITCZ) and is characterised by being the warmest area of the planet and can be recognised by the permanent clouds that can be observed all year round.

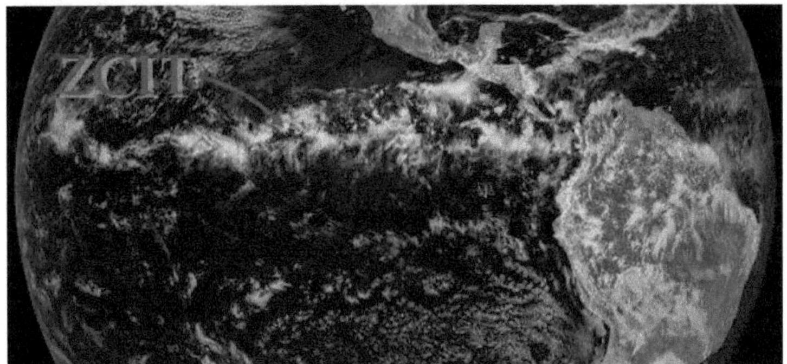

FIGURE 3.07. Source: http://upload.wikimedia.org/wikipedia/commons/thumb/1/12/IntertropicalConvergenceZone-EO.jpg/300px-IntertropicalConvergenceZone-EO.jpg

The air masses in the lower part of the circulation cells constitute the trade winds (red arrows in figure 3.06). These constant winds deviate their trajectory to the right in the northern hemisphere and to the left in the southern hemisphere, due to the Coriolis effect.

FIGURE 3.08. Source:
http://www.inamhi.gov.ec/educativa/diccionario/ letraA.htm

Those areas where the cells converge and the winds descend to near the surface are called anticyclones, which in the southern hemisphere rotate in a counterclockwise direction. These are areas of high pressure and are characterised by stable climates. In our continent, the South Pacific anticyclone is the one that governs the trade winds in most of South America, including Peru. In our case it is one of those responsible for the climate we have and for the peculiarities of our sea: the cold climate despite being in a tropical zone, a desërtic coast where there should be jungle (as is Brazil, which is at the same latitude), a sea rich in primary, secondary and tertiary productivity, etc.

Other winds go from the equator to the poles, in the opposite direction to the trade winds, and are therefore called contra trade winds. Another type of wind is the monsoon, characterised by a seasonal change of direction (as in the Indian Ocean). There are other, less significant types of wind, such as the general westerly winds, with a southwesterly direction in the northern hemisphere and a northeasterly direction in the southern hemisphere, which sometimes reach high speeds.

Ocean-atmosphere interaction

As we have seen, numerous atmospheric parameters and phenomena interact with the oceans. Many climatic and oceanographic phenomena have components in both environments, and it is often difficult to determine where one ends and the other begins. We will look at the main interacting phenomena that are important for the understanding of the oceans.

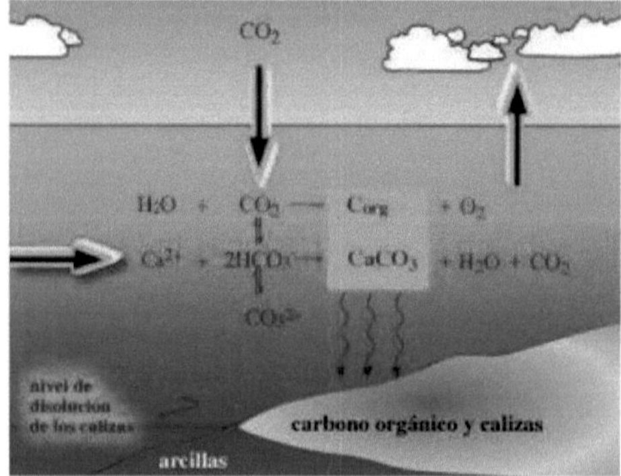

FIGURE 3.09. Source:
http://foro.meteored.com/climatologia/relacion+phco2+i+la
+constante+de+equilibrio+del+h2co3-t24897.0.html

The **chemical composition** of the ocean is influenced by the chemical composition of the atmosphere. Indeed, by diffusion processes and above all by mixing processes the

air enters the ocean increasing the concentration of dissolved gases such as oxygen, allowing the existence of aquatic animal life. Carbon dioxide is another gas that also enters the ocean by these means, although in much lower concentrations than in air, and reacts with water to form carboxylic acid, which in turn, if the concentration is adequate, will be converted into bicarbonates and in turn into carbonates. The carbon system described depends on the pH of the water and the concentration of carbon dioxide. This system is important in the balance of ions and chemical reactions for the support of aquatic life. Another gas that is exchanged between the ocean and the atmosphere is gaseous nitrogen, although the contribution to the ocean by this means is negligible under normal conditions.

These mixing processes involved in gas exchange are due to the turbulence generated by the wind, which can reach several metres deep (and can reach several tens of metres), homogenising the concentration of their chemical composition and other parameters such as temperature (also known as homotherm), and can reach saturation levels in these dissolved gases.

Another type of energy exchange between the atmosphere and the ocean is the force exerted by winds on the surface of the oceans and its main result is ocean currents and waves. Large ocean currents produce important changes in the climate of coastal areas and the planet as a whole. The atmospheric **current system** has a strong relationship with the general circulation of the oceans, as shown in Figures 10 and 11. This is because the main driving force of ocean surface currents are the winds, especially the permanent ones such as the trade winds (lower component of the cells, Figure 3.06).

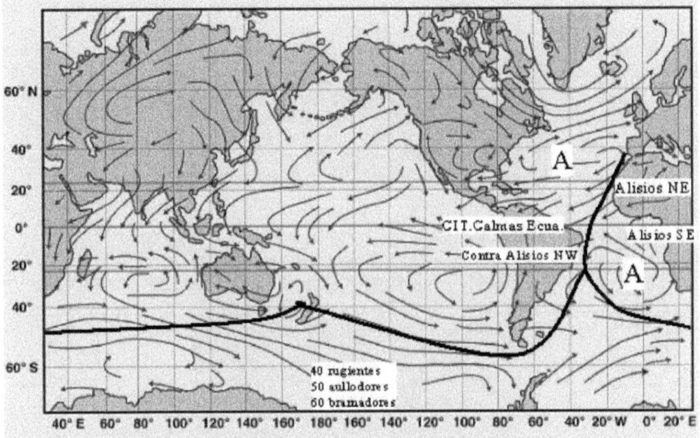

FIGURE 3.10. Source: **www.marviva.org/articulos/regatasmundo2.htm**

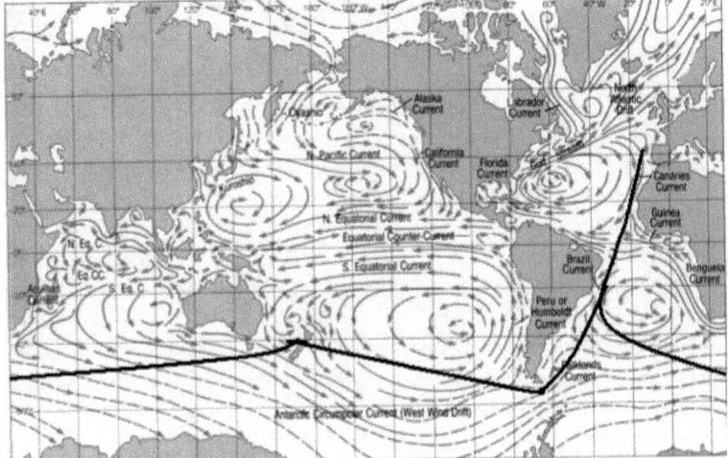

FIGURE 3.11. Source: **www.marviva.org/articulos/regatasmundo2.htm**

The circulation of the rotating winds is very similar to the circulation of the rotating water masses in each of the ocean basins of both hemispheres and in the same directions, clockwise in the northern hemisphere and counterclockwise in the southern hemisphere. Thus, we have the counterclockwise gyre in the Southern Pacific Ocean, of which our Peruvian current is part, located on the eastern side of this gyre. This movement of the sea is fundamental to sustain life on Earth.

These wind manifestations can be violent and cause great damage.

When an invasion of cold air from high altitudes occurs, displacing warm air from the lower layers of the atmosphere, so-called tornadoes are produced in which the cold air mass descends with a spiral motion and with an intensity of up to 350 km/h. Tornadoes occur mainly in North America.

FIGURE 3.12. Source:
http://www.fotos.org/galeria/data/538/3tormenta- tropical-photos-from-space.jpg

FIGURE3.13.Source: http://www.pixelicia.com/ wp-content/uploads/2007/10/tormentajpg

Wind speed causes evaporation at the sea surface to intensify rapidly, increasing the energy in the atmosphere and generating storms; as storms progress, evaporation increases, supplying more energy to generate more storms. This process of energy regeneration is one of the factors in the formation of the catastrophic tropical storms called hurricanes in the Atlantic Ocean and typhoons in the Pacific Ocean. These storms continue to increase in strength while in the temperate ocean, and decrease in strength as they pass over land or areas where the ocean is cooler and evaporation is reduced. It is also customary to call these violent wind manifestations cyclones.

On the other hand, non-permanent winds are responsible for another type of ocean water movement: **waves**. These movements are due to near-surface winds that move the water in a cylinder without involving displacement of the water mass. The water molecule is the one that rotates in a circular shape (see

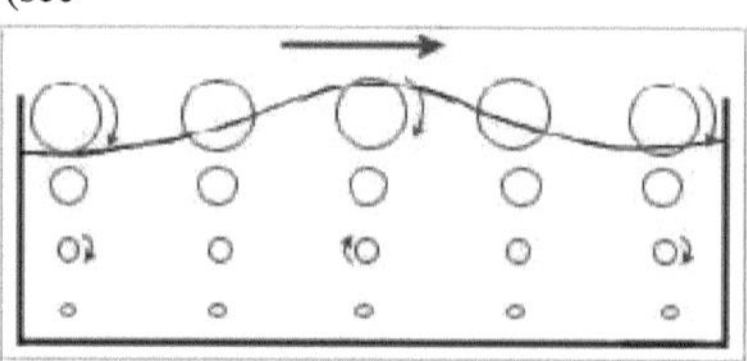

FIGURE 3.14. Source:
http://www.contraola.com/2008/01/qu-son-las-olas- un-poco-de-teora.html

figure 3.14) while maintaining its position, but drives the motion of adjacent molecules, causing it to move in a circle as well. This is how the wave moves forward in the ocean.

In other words, during the passage of a wave, each water molecule undergoes a more or less circular motion, first upwards and forwards and then downwards and backwards. At a given moment the surface molecule will be at the bottom, i.e. the valley, and when it is at the top it is called the crest (see figure 3.15). When the wave reaches the coast and the cylinder rubs on the

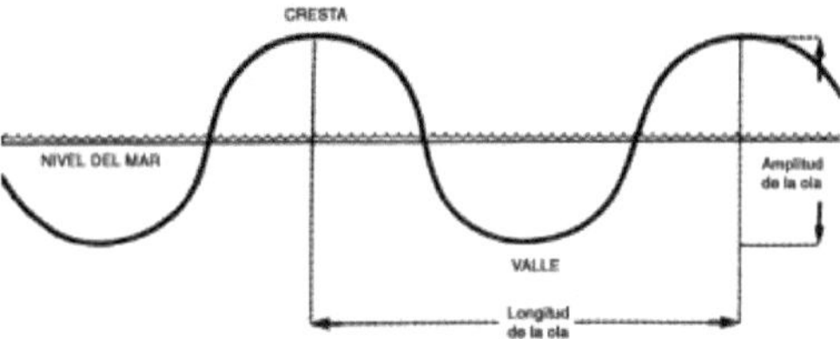

FIGURE3.15.Source:http://www.profesorenlinea.cl/swf/links/frame_top.ph p?dest=http%3A///www.profesorenlinea.cl/geografiagral/Hidrosfera.htm

The wave is then rolled over the bottom of the water mass and the wave breaks. This sequence can be seen in figure 3.16.

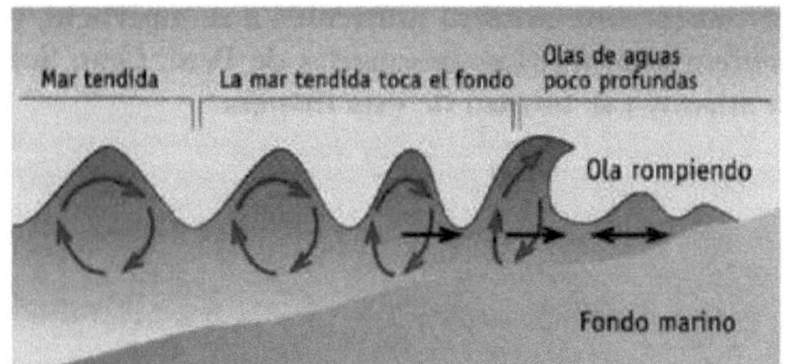

FIGURE3.16.Source:http://http://www.profesorenlinea.cl/swf/links/frame_top.ph.profesorenlinea.cl/swf/links/frame_top.ph p?dest=http%3A///www.www.profesorenlinea.cl/geografiagral/Hidrosfera.htm.cl/geografiagral/Hidrosfera.htm

The heat exchange between the ocean and the atmosphere is a real thermal engine. This exchange causes the poles and the equator to have different temperatures. This differential heating in turn causes the circulation of air masses in the atmosphere and water masses in the oceans, which results in the temperature remaining more or less constant in the different regions of the Earth.

In the regions between 20° and 40° latitude, the radiation arrives without any loss and is converted into heat energy. Therefore, if only the effect of solar radiation were taken into account, there would be a constant change in the Earth's temperature; the tropics would tend to become warmer, and the polar regions would become colder and colder. Fortunately, the Earth's temperature does not depend on radiation alone, since there is a flow of energy from the warm tropics to the poles that is carried by the atmosphere as well as the oceans.

Taking into account the temperature of the equatorial and polar zones, the average ocean water temperature is 3.8°C. In the southern hemisphere the surface is 1°C warmer than in the northern hemisphere; however, temperatures in the southern hemisphere, for any latitude, are generally lower than those corresponding to the same latitude in the northern hemisphere.

The solar radiation that passes directly through the atmosphere until it reaches the surface of the planet is absorbed by the continents and the oceans; as the latter occupy 70% of the globe, most of the energy coming from the Sun is stored on the surface of the sea and it is this absorbed radiation that heats the atmosphere, first in the atmosphere and then in the oceans.

The heat is transported from the tropics to the higher latitudes at the poles, where it radiates its energy into space.

Another form of heat exchange between the ocean and the atmosphere results from the evaporation of water from the ocean surface, which produces heat in the amount of 600 calories per gram of water vapour. This vapour, once it leaves the surface of the sea, rises through the free air until it condenses into rain, generating thunderstorms. To illustrate the importance of the energy produced by the evaporation of ocean water, it is worth noting that the energy released by the condensation of this vapour is equal to the electrical energy used by 10 cities the size of New York.

The relationship between evaporation and precipitation in the oceanic and atmospheric regions is very important for heat transfer on the planet. In summary, the forms of energy transfer from the oceans to the atmosphere are, according to their importance: radiation from the sun, evaporation, thermal energy exchange by heating or cooling the air, and mechanical energy exchange caused by atmospheric pressures and winds.

CHAPTER IV

HEAT BALANCE

The ocëano is a mass of water which is generally cold. Only the thin surface layer is heated by the Sun, which is the main source of our planet's energy, which comes in the form of electromagnetic radiation, although it behaves both as a wave and as a particle (which is called a photon).

Electromagnetic radiation from the sun, or light, travels in the vacuum at 299792 km.s-1 in a range of frequencies that are distinguished by their different wavelengths (λ). λ λ Some, such as radio waves, have wavelengths as long as kilometres, while the most energetic (λ short and high frequencies), such as gamma radiation or X-rays, have wavelengths as short as nanometres (nm), which penetrate very thin materials with some ease, which is why X-rays make it possible to observe the inside of the human body.

The energy reaching the outside of the atmosphere is a fixed quantity, called the -solar constantl. Its value is 1400 w/m2, which means that 1 m2 located in the outer part of the atmosphere, perpendicular to the line that joins the Earth to the Sun, receives 1400 J every second. It is a mixture of radiation of λ between 200 and 4000 nm. A distinction is made between ultraviolet radiation, visible light and infrared radiation.

Ultraviolet radiation. From lower λ

of 360 nm, carries a lot of energy and can disrupt molecular bonds. Those smaller than 300 nm can affect molecules such as DNA and would cause irreparable damage if it were not for the fact that they are absorbed by the

---

The variability of the solar constant

By measuring its variability in space and time over the Earth, the basic radiative forcing of the climate system can be defined. This value gives an idea of the values that are recorded in the upper atmosphere and of the values that finally reach the earth's surface during the day as a consequence of radiation losses due to phenomena (attenuation processes) such as reflection, refraction and diffraction during its reflection, refraction and diffraction during the day:

trajectory.

I_0 = 1.367 W/m^2

= 433.3 Btu/(ft^2 *h)

= 1.96 cal/(cm^2 *min)

Visible light. The radiation in the visible zone whose A is between 360 nm (violet) and 760 nm (red). Because of the energy it carries, it has a great influence on living beings. This light passes through a clear atmosphere quite efficiently, but when there are clouds or dust masses, part of it is absorbed or reflected.

Infrared radiation. Of more than 760 nm, it is the one that corresponds to longer wavelengths and has little energy associated with it. Its effect is to accelerate the reactions or increase the agitation of the molecules called heat by increasing the temperature, it does not affect the bonds of the molecules. CO_2, water vapour and the small water droplets that form clouds absorb these radiations very intensely.

In addition to the Sun, other sources of energy are the **internal energy of the Earth** whose inner core reaches 5000°C1 and is responsible for the convection currents that move the lithospheric plates, thus having important repercussions on many surface processes: volcanoes, earthquakes, movement of continents, formation of mountains, etc. Another source of energy is **cosmic radiation**, with very short A's that reach the upper atmosphere as primary radiation (formed by high-energy electrons), which hits the molecules and becomes secondary radiation (ultraviolet rays). The oxygen molecules (O_2) absorb the primary and secondary radiation of less than 200 nm and are converted into ozone (O_3). Ozone in turn absorbs radiation up to 300 nm, and thus, thanks to oxygen and ozone, the Earth is protected against the most dangerous cosmic radiation. Finally, another source of energy is **radioactive substances** of much lower magnitude, consisting of a group of short-wave radiations, which are therefore very energetic and have a high penetration capacity. Their origin may be natural, but they have increased in recent years as a result of certain human activities, in particular nuclear explosions.

[1] The energy source that maintains these temperatures is mainly the radioactive decay of chemical elements in the mantle.

4.1. Calculation of heat in the earth

The calculation of heat on earth is very simple. All the heat comes from the sun and an equal amount is returned by the earth to space. That is why the earth's temperature remains constant at about 15°C.

While the energy received is a mixture of ultraviolet, visible and infrared radiation, the energy returned by the Earth is mainly infrared and some visible. This difference is due to the fact that the radiation arriving from the sun comes from a body that is at 6000°C, but the radiation returned from the surface has the wavelength composition of a black body that is 15°C. For this reason, the reflected radiations have wavelengths of lower frequency than those received.

The total heat dynamics of the Earth involves: the sea-air interaction zone, the energy from space that passes through the atmosphere and is absorbed by the ocean, the oceans that heat the overlying atmosphere, and the atmosphere that transports the energy to the polar regions, where it is emitted into space in the form of radiation.

Looking at the heat flows within the planet in detail, we find differences. At low latitudes the energy input from the Sun is greater than the energy lost to space by radiation; at high altitudes, on the other hand, the energy input from the Sun is less than the energy lost to space. Most of the sunlight passes through the atmosphere without being absorbed because the air is almost transparent to the various wavelengths.

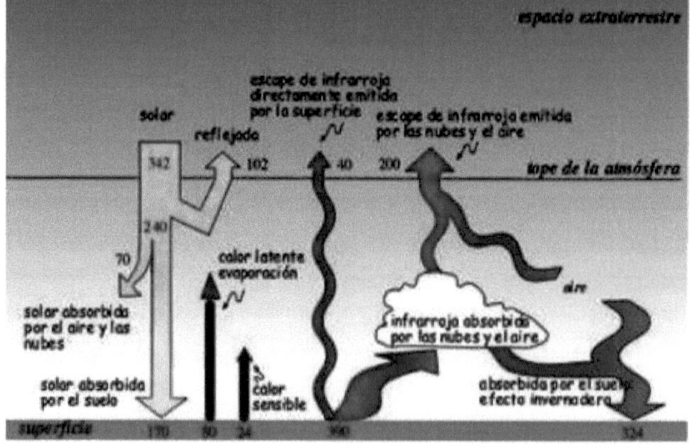

FIGURE 4.01.Source: *www.ideam.gov.co/radiacion.htm*

emitted by the sun, but when sunlight hits the surface it is absorbed and converted into heat. The heated surface transmits its heat to the atmosphere, partly by radiation, partly by heating the cooler air and partly by 1атЫёп in the form of latent heat of vaporization, because the heat extracted from the water, when evaporation occurs, is transmitted to the air at the time when the water vapour condenses in the form of clouds and rain. The heat loss to space depends mainly on the radiation from the atmosphere.

Under conditions of a perfectly clear day and with the sun's rays falling perpendicularly, almost three quarters of the energy arriving from outside reaches the surface. Of the latter, 49% is infrared radiation, 42% visible light and 9% ultraviolet radiation. On a cloudy day a much higher percentage of energy is absorbed, especially in the infrared region. Another part of the energy is absorbed by vegetation in the whole spectrum, especially in the visible light area, which is used for photosmethylation.

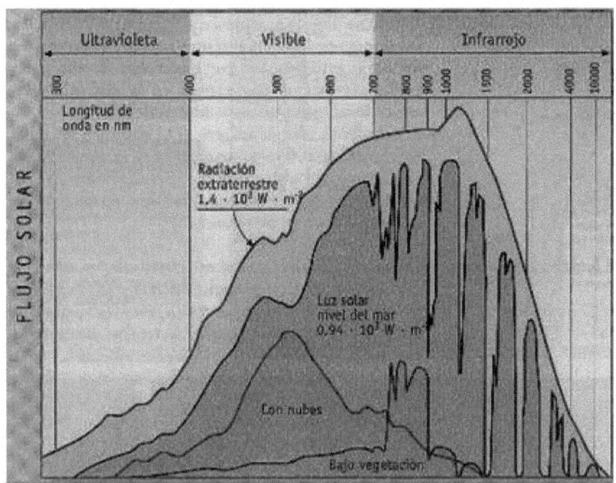

Figure 4.2. Distribution of solar radiation in the upper atmosphere and at sea level, under different circumstances. Source: http://www.tecnun.es/
Subjects/ecology/Hypertext/02Earth/110BalEner.htm

Referring to energy, the amount that reaches the upper atmosphere is constant (known as the solar constant) whose value is 1400 W/m^2 (more exactly 1368) and since the surface of the earth is four times larger than the area of the visible circle of the sun at any time, the average amount of solar radiation received by the earth's surface is 1368/4 = 342 W/m^2 . Whereas this amount constitutes 100% of the solar radiation received by our planet (Figure 4.1), 30% (102) is lost to space, either reflected from the Earth's surface (4%), clouds (20%) and dissipated by the atmosphere's own heat (6%). The remaining energy allows the Earth to be heated and is distributed as follows: 20% (70) is absorbed by the atmosphere (16% is absorbed by gases and dust and 4% is absorbed by clouds) and 50% is absorbed by the surface by both direct radiation (21%) and indirect 25% (6% by dissipation from the heated atmosphere at the surface and 19% is radiation reflected by the cloud).

Table 1. Radiation balance in W/m^2

ENTRANCE		OUTGOING	
Heat balance of the earth's surface			
Solar radiation	170	Terrestrial radiation	390

Atmospheric radiation	324	Evaporation	80
		Conductivity and Convection	24
Total	**494**	**Total**	**494**
Atmospheric heat balance			
Solar radiation	70	Radiation to space	200
Condensation	80	Radiation to the surface	324
Terrestrial radiation	390	Radiation from earth to space	40
Driving	24		
Total	**564**	**Total**	**564**
Planetary heat balance			
Solar radiation	342	Reflected and dispersed	102
		Radiation of atmosphere and clouds into space	200
		Radiation from earth to space	40
Total	**342**	**Total342**	

4.2. Surface heat calculation

The calculation of surface heat can be expressed as follows: Radiation received from the sun = radiation leaving the earth + sensible heat transmitted to the air by conduction + heat used in the evaporation of water. In symbols we have:

$$Qs = Qb + Qh + Qe$$

This equation is applicable on a large scale and at sufficiently long time intervals so that surface temperature changes can be neglected. If the surface temperature is changing, a term Qt is added to represent the heat that causes the temperature to rise. And if the surface is fluid, as are the oceans, any heat extracted out of or into a region by currents must be represented by a tëterm Q_v. Thus, for a short period of time and a particular locality the equation is:

$$Qs = Qb + Qh + Qe + Qt + Qv$$

Figure 4.2 represents the average conditions over the earth as a whole, for both the earth's surface and the oceans. Only over the oceans is the përadiative heat loss lower than that over the earth, and all other terms are higher, especially those relating to evaporation. The calculation of the heat of the oceans is estimated as follows:

$$Qs = Qb + Qh + Qe$$

Shortwave radiation from the Sun, most of which is in the visible spectrum, is dissipated both by absorption by the atmosphere at high altitude and by water vapour at low levels. In other words, the amount of sunlight that reaches the earth's surface depends mainly on cloud cover and the angle of the sun. Cloud type is very important, because thick clouds reduce energy much more than high thin clouds. The average height of the sun determines to a large extent the amount of radiation received. The energy reaching the earth is greater at lower latitudes than at higher latitudes. The angle of the sun also affects the absorption of the earth's surface because with a low sun, more light is reflected from the surface than is absorbed.

4.3. The Global Ocean Heat Balance

Water absorbs more of the incident radiation than land surfaces. As a rough average, water absorbs 94% of the incident energy, while land surfaces vary between 10 - 20% reflection, while the reflection of snow or clouds can be as high as 80%. Thus, the oceans absorb large amounts of energy, and hold a huge reservoir of it in the form of heat.

The ocean heat balance is composed of inputs and outputs. A complete list of all inputs and outputs is given below; where - +‖ denotes heat input or heat gain and - -‖ denotes heat output or heat loss.

Main inputs and outputs

- solar radiation (+)
- longwave back radiation (-)
- direct air/water heat transfer (sensible heat transfer) (-; + when from air to water)
- evaporation heat transfer (-; + for condensation which rarely occurs)
- advection heat transfer (currents, vertical convection, turbulence) (**-or** +); this effect is cancelled out on a global scale or in closed basins.

Secondary sources:

- heat gain from chemical/biological processes (+)
- heat gain from within the Earth's interior and from hydrothermal activity (+)
- heat gain from friction currents (+)
- heat gain from radioactivity (+)

The contributions from secondary sources are negligible in most cases, so only the primary inputs and outputs will be taken into account in this paper.

Incoming heat: Solar radiation. As already mentioned, at the upper edge of the atmosphere there is 1400 W/m^2 The solar constant varies seasonally between 0 and 530 W/m^2 at the poles and between 390 - 440 W/m^2 at the equator. The maximum interannual variations arise from the variation of the distance between the Earth and the Sun and reach 3.34%[2] .

Татыёп pointed out that not all the radiation received from the outermost Hmite of the atmosphere is available in the oceans. It varies irregularly with wavelength, as a result of absorption by water vapour and various atmospheric gases, in particular oxygen and hydrocarbons. Absorption in the sea decreases rapidly with depth (Figure 4.3 in the box below). Table 4.2 shows in percentages the amount of light reaching different depths of the sea (under favourable conditions):

Table 2. Percentage of light penetrating the sea at different depths

Light	Penetration	
73,0%	Reach	1 cm deep
44,5%	Reach	1 m deep
22,2%	Reach	10 m deep
0,53%	Reach	100 m depth
0,0062%	Reach	200 m depth

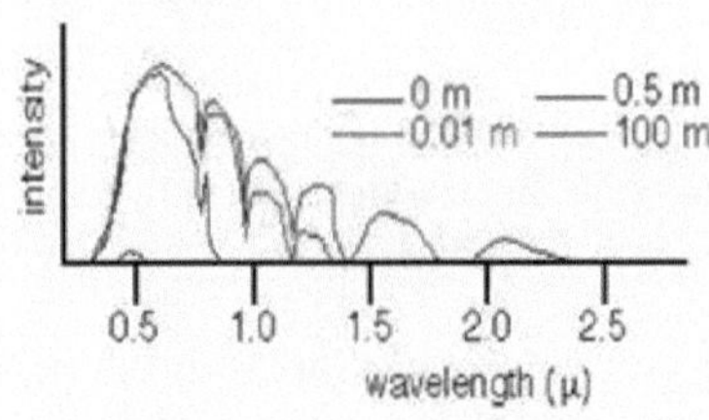

Figure 4.3. Spectral distribution of solar radiation received at different depths as a function of wavelength. The different minima in the incoming intensity at sea level (0 m) are caused by the absorption of atmospheric gases

---

[2] Some literature still uses the unit cal cm^{-2} dla^{-1} (calories per square centimetre per dla) but nowadays it has been replaced by the unit W m^{-2} (Watts per square metre). 1 cal cm^{-2} dla^{-1} =0.484 Wm^{-2} .

(mainly water vapour, carbon dioxide and ozone). (wavelength: wavelength; *intensity*: intensity)
Source:www.es.flinders.edu.au/~mattom/IntroOc/curso04. html

So, most of the energy is captured in the shallowest layers of water, which are therefore the most effectively heated. The portion of energy in the visible spectrum penetrates further, but attenuates rapidly with depth. Of course, it also depends on the degree of turbidity of the water, the higher the turbidity the lower the penetration.

In addition, the minimum amount of energy needed to maintain photosmethesis is 0.0015 W/m^2 . Under optimal conditions (completely clear water) this amount is available down to 220 m depth.

This energy from the sun heats the surface water and penetrates downwards due to wind and wave turbulence. Since the warm surface layer is lighter in weight than the cooler waters, mixing takes place at a depth determined mainly by the wind force. A distinctly pronounced transition often develops between the well-mixed warm surface layer and the layer of cooler water below it. This transition layer is called a thermocline layer.

A. *Outgoing heat: Return radiation*. The surface of the ocean not only absorbs shortwave solar radiation, but some of it radiates back to the atmosphere (water vapour and clouds absorb almost all of the heat, and radiate much of it back to the surface)[3] and a small part escapes directly into space as longwave radiation. The wavelength at which most of the back radiation occurs is explained by Wien's Law (see box below). As the sea surface temperature is lower than that of the Sun (~283 K), the maximum of the back radiation is located at about 10 microns, i.e. in the infrared.

According to Stefan-Boltzman's law, the energy of radiation is proportional to the fourth power of its absolute temperature (°K). Thus, daily seasonal variations in the surface temperature of the ocean have little effect on the return radiation energy, because these variations are small compared to the absolute temperature level.

Wien's law of displacement

As the temperature of a body increases, the maximum of its energy distribution shifts towards shorter Л, causing a change in the colour of the body. This law is very useful for determining the temperature of hot bodies, such as furnaces or stars, by finding the Л for which the intensity of radiation is at its maximum. For example, at a temperature of 200°K a body emits visible light but the intensity at the red (low frequency, high Л) end of the visible spectrum is much greater than the blue (high frequency, low Л) and the body appears bright red. At 3000°K, the approximate temperature of an incandescent lamp filament, the relative amount of blue light has increased, but the red component still predominates. At 6500°K, which is the temperature of the Sun, the distribution is almost uniform among all components of visible light and the body appears bright white. Above 10000°K blue light is emitted more intensely than red and a body (hot star) at this temperature appears blue.

On average, incoming radiation exceeds outgoing radiation at all latitudes. The excess amounts to about 0.082 W/m^2 in the tropics, and about 0.019 W/m^2 in the 60 - 70° latitudes. The energy received must equal the energy lost, so the excess radiation received by the surface must be emitted in some other way. Part of this excess goes into contact heating of the air, when the air is cooler than the surface. This is called "sensible heat" because it is perceptible to the touch. A larger proportion of the excess is given off when water evaporates.

B.- *Outgoing heat. Direct (sensible) heat transfer between the ocean and the atmosphere.*

---

[3] The air over the oceans is normally more humid than the air over land, so much of the heat that escapes is captured and returned to the surface. The heat that escapes from the surface by radiation (back radiation) is less over the oceans than over land, is somewhat less over warm water than over cold water, and is more constant from day to night and between seasons than the fluctuating values of sunlight.

28

On average, the surface of the ocean is about 0.8°C warmer than the air above it. Therefore, direct heat transfer (sensible heat transfer) takes place from the water to the air, constituting a heat loss, in magnitude directly proportional to the temperature difference between the two media. Heat transfer from the ocean to the atmosphere is achieved much more easily than in the opposite direction for two reasons:

1. It takes much less energy to heat air than water. The eeiergni required to raise the temperature of a 1 cm thick layer of water by 1°C is enough to raise the temperature of a 31 m layer of air by the same amount.

Effect of back radiation on ice
formation

Water absorbs almost all of the incoming energy, but ice reflects more than half. The loss of heat through counter-radiation continues to cause an immediate radiation deficit. The result is a rapid drop in the temperature of the ice, and a rapid increase in the thickness and extent of the ice cover. The ice temperature drops until the outgoing radiation, reduced to lower temperatures, is again balanced by the reduced incoming energy. In the same way, when an ice sheet begins to break up, it can quickly disappear due to the increased heat absorption in the water.

2. Heat input into the atmosphere from below causes instabilities (by reducing the density at the base) resulting in turbulent upward heat transport. Conversely, heat input to the oceans from above increases the stability of the water column (by reducing the density at the surface) preventing efficient heat input to the deeper layers.

(What are the consequences of sensible heat exchange on the ocean and the atmosphere? When the surface is warmer, the heat moves upwards, and as in this case the air is heated from below, vertical currents are formed which

carry the heat to high levels. The overturning of the air keeps the renewal of cold air close to the surface, so that heat conduction from the warm surface continues rapidly. If the temperature difference between the surface and the air is very large, the turbulent overturning of the air can be so intense that it creates violent thunderstorms.

The most intense loss of sensible surface heat occurs off the eastern coasts of mid-latitude continents, where large warm water currents such as the Gulf Stream and the Kuroshio Current are invaded during the winter by large and violent turbulence of cold air from the snow-covered continents.

In summer, large areas of the ocean are cooler than the air above them. When the air is warmer than the water, the heat passes downwards by conduction, but in this case the cooling of the air at low levels increases its stability. There is little air exchange near the surface, and temperature equilibrium is reached quickly. Therefore, appreciable sensible heat exchange between water and air occurs only when the water is warmer.

C.- *Outgoing heat. Heat transfer by evaporation and condensation*. 50% of the heat input to the oceans is used for evaporation. In addition to the important contribution to the heat balance, evaporation (constituting a loss of water to the atmosphere) plays an important role in the mass balance.

Evaporation begins when the air is unsaturated with moisture. Warm air can hold much more moisture than cold air. Since under normal conditions the air is cooler than the sea surface, air that is unsaturated with moisture will be heated when it comes into contact with the sea, thereby increasing its moisture capacity, and evaporation occurs; the heat required to convert the rich water into water vapour is extracted from the sea. This heat of vaporisation, which is

29

lost at the surface when evaporation occurs, is considerably greater than the sensible heat loss, although the relative proportions vary. On a global average, the amount of heat lost by conduction is about 1/10 of that lost by evaporation, but the sensible heat loss is greatest when the air is very cold compared to that at the surface.

If the air is warmer than the sea and relatively dry, evaporation will take place only when the vapour content of the air equals the maximum that the air can hold at the sea surface temperature. If the warm air is already more humid, the opposite process of condensation will take place over a small area of the surface, and the heat of evaporation released to the sea in this case is negligible. What is more important is that the warm, moist air is cooled by contact with the sea to a point lower than the dew point; that is, saturation temperature and the condensed excess moisture is converted into fog. Like sensible heat loss, evaporation is most intense in winter over the large warm currents that move poleward off the eastern coasts of the continents.

The subtropical trade wind zones are also regions of high evaporation, with little seasonal variation. Generally, clear skies and high overhead sunshine combine to produce a large radiation surplus, most of which is converted into evaporated water. Due to the large extent of the Octians at these latitudes, the greatest volume of water is lost in these subtropical regions.

Condensation occurs when warm air meets cold water. The oceanic areas where this occurs are known and feared for the frequent occurrence of fog. Most of the energy released during condensation goes into the atmosphere, so the contribution of condensation to the oceanic heat balance is extremely small.

In summary, the heat balance in octians is the adjustment between the terms discussed above. Normally, the first two terms are not considered separately; the difference solar radiation minus return radiation, or net radiative heat gain, is used as the major input. The balance is then

net radiative heat gain - evaporative heat loss - direct colour loss = 0

This adjustment works if all the world's oceans are considered. If the balance is evaluated for limited regions, the right-hand side of the equality is generally not zero, but represents the heat transfer achieved by ocean currents.

On a global scale, Figure 4.4 illustrates the role of heat transport by currents in the heat balance in a zonal section from 60°N to 60°S. The net heat input decreases from the tropics towards the poles; a weak decrease occurs at the equator in view of the thick cloud cover in the area. Maximum evaporative heat loss occurs in the subtropics due to atmospheric advection of dry air; the decrease in the tropics is the result of the high moisture content of tropical air. Direct heat loss is small over the whole area. The currents take heat from the tropics (a heat loss for that portion of the ocean - positive values) and deposit it in the subpolar regions (a heat gain - negative values).

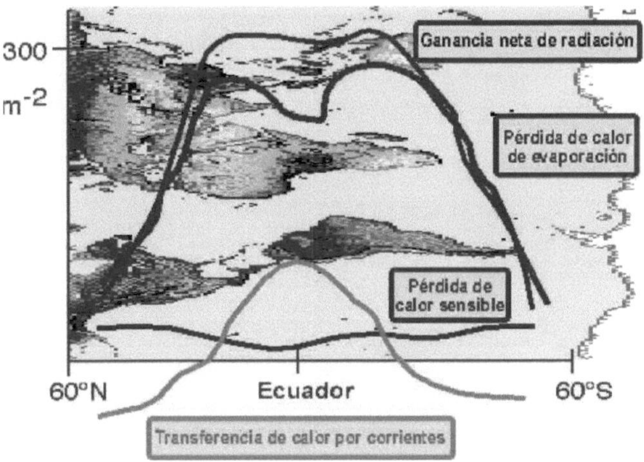

Main contributors to the heat balance

Figure 4.4. A qualitative plot of the important components of the heat balance, averaged zonally over the oceans, as a function of latitude.

4.4. Effects of planetary heat imbalance: GREENHOUSE EFFECT

The greenhouse effect, often confused with global warming, is a natural phenomenon that allows the Earth to maintain a temperature suitable for habitation. Some gases in the atmosphere retain part of the energy emitted by our planet's soil and obtained from solar radiation. These gases, such as carbon dioxide, methane, nitrous oxide and ozone, which are known as greenhouse gases, are part of the natural composition of the Earth's atmosphere.

Since the industrial revolution and especially since the second half of the 20th century, human activity (such as the burning of fossil fuels [coal, oil, natural gas], deforestation or the creation of landfills) has increased the amount of these gases in the atmosphere, leading to an imbalance in the Earth's temperature, which causes global warming.

Although in various international forums it is confirmed that the industrialised countries are mainly responsible for global warming, according to the United Nations Framework Convention on Climate Change, the five nations that produce the greatest amount of carbon dioxide, the main cause of climate change, are the United States (39.4%), Russia (5.9%), Japan (4.9%), Germany (3.2%) and Canada (2.3%). However, it must be recognised that the entire world's population contributes daily to raising the temperature of the planet.

Figure 4.5. Source: www.foroambiente.blogspot.com/

CHAPTER V

OCEAN VARIABLES

The study of the oceans is of great importance to modern globalised society because of the strong interrelationships between different communities around the world. The science of oceanography has become so important that in the last three decades it has received billions of dollars in funding to continue the understanding of the oceans.

The progress achieved in the field of oceanography is remarkable, mathematical models have been established that predict some phenomena and behaviours with high accuracy for periods up to three months into the future.

The parameters most inherent to the knowledge of the oceans, which cannot be understood without recourse to these are known as variables. These can be grouped into those that can give exact information such as temperature, salinity and pressure; and those that can be approximated such as turbulence and ocean current.

In this chapter we will only focus on the variables temperature, salinity and pressure.

5.1. Temperature

Parameter that measures the degree of heat of a body. It is defined as the measure of the kinetic energy of molecules in motion. It is commonly expressed in degrees Celsius (°C). The sun is the most important source of heat for the oceans. It is the easiest parameter to obtain in the sea. Due to the structure of the water molecule, the oceans have a high heat capacity, high latent heat of fusion, high latent heat of evaporation. Seawater is also a good conductor of heat, compared to other fluids.

The temperature of the oceans lies between -2° and 30°C. The lower Hmite is determined by the formation of sea ice, and the upper Hmite is determined by the processes of radiation and heat exchange with the atmosphere (in areas close to land the temperature may be higher, but in the open ocean it rarely exceeds 30°C). About 90% of the volume of the world's ocean has a smaller range, ranging from -2°C to 10°C. The temperature at the seafloor is always low, ranging from 4°C to -1°C, so that high pressures are associated with low temperatures.

The distribution and variation of temperature is closely linked to currents and variations in solar radiation. The Pacific Ocean is the warmest of the oceans (19.37°C), followed by the Indian Ocean (17.27°C) and the Atlantic Ocean (16.53°C). Temperatures in the Northern Hemisphere are on average 3°C warmer than those in the Southern Hemisphere at all latitudes, due to the configuration of the Océanos, the current system and the influence of the poles. In the southern hemisphere all three océanos are completely open and therefore under the influence of Antarctica. In the Pacific Ocean the western half of the tropics is warmer than the eastern half, which is important for the temperature distribution and the development of the El Nino phenomenon.

In general, annual mean temperatures give isotherms with a latitudinal distribution. They are higher at the equator and decrease towards the poles, i.e. the isotherms have a zonal distribution (Figure 5.01). In all oceans the highest surface temperature values are found slightly north of the equator, which is related to the Intertropical Convergence Zone (linked to the atmospheric circulation in the two hemispheres). On the other hand, the South Pacific anticyclone deflects isotherms northward at high latitudes and westward near the equator, as is the case off the coasts of Chile, Peru and Ecuador; off the Peruvian coast the upwelling further increases the zonal temperature gradient.

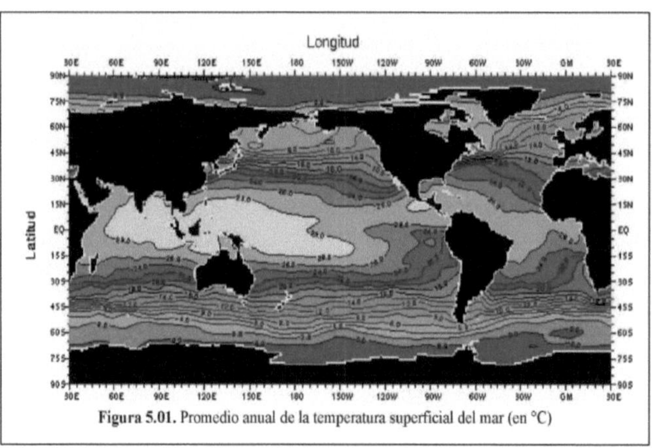

Figura 5.01. Promedio anual de la temperatura superficial del mar (en °C)

Isotherms from the equator to the poles have made it possible to divide the ocëano into zones. Up to the 20°C isotherm, the hot (tropical) zone can be defined, between the 20°C and 12°C isotherms the subtropical zone, between the 12°C and 6°C isotherms the temperate or subpolar (subantarctic or subarctic) zone, and below the 6°C isotherm the polar (antarctic or arctic) zone. Between the latitudes of 75° and 80° (north and south), the average temperatures drop to -1°C and -1.8°C. This division has some anomalies due to the circulation of waters that transfer energy in the meridian direction. Thus, in the eastern end of the tropical zone the waters are colder than in the west because the oceanic circulation moves warm subtropical waters towards the equator, which in turn generates the upwelling phenomenon. For this reason, the temperate zones have a west-east tërmic asymmetry.

Periodic and aperiodic, diurnal, seasonal, annual and decadal surface temperature variations have also been recorded, always less pronounced than those of the atmosphere. Offshore, the diurnal amplitude is of the order of 1°C, while the seasonal oscillation depends strongly on latitude: strong in the temperate zone (6 to 9°C) and weaker in the warm and cold zones (4.5°C). Near the coasts and in coastal seas, the annual amplitudes increase.

With respect to vertical distribution, from the surface to 2000 m depth the temperature decreases rapidly in the tropical and subtropical zones, while in the polar zones the decrease is negligible or even occurs as an inversion phenomenon where the surface waters may be colder than in the underlying layers. From 1500 to 2000 m the temperature decreases very slowly and overall the oceans tend to become more uniform in the 0 to 3°C range. At certain depths, especially in the upper layer, there is an intensification of the so-called thermocline gradient, such as that between surface waters of varying temperature and the underlying waters of more stable temperature.

Depending on the region and the season, this thermocline can be found at a depth of a few tens of metres or up to 600 - 700 m depth. Naturally, diurnal and seasonal oscillations disappear rapidly at about 20 m for surface waters and at about 100-150 m for underlying waters. In general terms, from a thermal point of view, it can be said that the ocean consists of a warm part, above 10°C, which extends between the surface and 500 m depth and does not reach beyond 50°N and 45°S, and which is overlain by a cold part (temperature below 10°C), which is much more voluminous, since it occupies all the remaining part of the ocean between 500 m and the bottom, and which - outcrops on the surface north and south of 50°N and 45°S;

It is the only part of the oceanic waters where ice forms.

In practice, temperature measurements in the sea (*in situ*) are made with specially designed mercury thermometers (reversible thermometer) that record the temperature when a sample of seawater is taken at a depth or with an electrical resistance thermometer.

The bottles commonly used for seawater sampling, each containing a thermometer protected from the effects of pressure by a glass envelope, have an accuracy of ±0,01°C. Two corrections must be applied to the readings of each thermometer: The first is to correct for instrumental error and is provided by the manufacturer, for each thermometer, and; since the temperature and pressure at which the thermometers are read are different from those at which they were taken, the second is to correct for instrumental error and is provided by the manufacturer, for each thermometer, and; the third is to correct for instrumental error.

The sample, a second correction must be applied by expansion of the glass and mercury. Finally they are compared with angle and wire length data.

Sampling depth can be determined using cable length and angle only if the angle is less than 5° and the cable length is several hundred metres. The depth of deep samples can be determined by indirect means.

With a knowledge of sampling depths, plots of sampling depths or plots of parameter distributions can be prepared, either for each station or for a series of stations. The preparation of a T-S diagram, the construction of which is explained below, can help not only in the identification of water bodies but also in the identification of errors that may contaminate the data.

Annual temperature variation

The surface area in any given region depends on many factors, among the most important of which is the variation during the year of the solar radiation, the behaviour of currents and prevailing winds. The characteristics of the annual variation of the surface temperature changes from place to place. Generally speaking, the annual surface temperature range is greater in the North Pacific and North Atlantic Ocean than in their southern parts.

> ### *Standard depths*
>
> When oceanographic data are published, the original dataset and also the data set interpolated to the standard depths which are, by agreement of the *Association, to* be included.
>
> *International Physical Oceanography*, 0; 10; 20; 20; 30; 50; 75; 100; 150; 200; (250); 300; 400; 500; 600; (700); 800; 1000; 1200; 1500; 2000; 2500; 3000; 4000 metres... up to the sampling limit. The graphs of the temperature distributions are shown in the following table.
>
> The values of these parameters at standard depths can help us to accurately estimate the values of these parameters at standard depths.
>
> The publication of the data in standard form facilitates comparative studies. This is followed by the calculation of parameters that are functions of the mean parameters, such as density, potential temperature, dynamic height, etc.

In the subsurface layers, temperature variation depends on four factors: variation in the amount of heat that is absorbed directly at different depths; the effect of heat conduction; variation in currents related to the vertical displacement of water masses, and; the effect of vertical movement (such as upwelling).

5.2. Salinity

As already indicated, seawater contains 3.5% salts, dissolved gases, organic substances and particulate matter in suspension, and their presence influences, to some extent, most of the physical properties of seawater, such as density, freezing point, temperature of maximum density (figure 5.2), electrical conductivity, etc., but does not develop new properties. Some properties that are not significantly affected by salinity are viscosity, light absorption, compressibility, thermal expansion and refractivity. Two properties that are determined by the amount of salt in the sea are conductivity and osmotic pressure.

One effect of salinity is that it raises the values of some thermal properties of the sea; for example, the increase of specific heat, thus ocean currents carry a lot of thermal energy; the high latent heat of fusion causes the temperature in polar regions to remain close to the liquefaction point; the high latent heat of evaporation is important in the transfer of heat from the sea to the air.

Salinity is conventionally defined as the total amount, in grams, of solid matter dissolved in one kilogram of water, all carbonates having been transformed into oxide, bromine and iodine replaced by chlorine, and all organic matter oxidised, all drying at a temperature of 480°C. The determination of salinity in this way requires the application of a very refined technique which is not practical in applied and common cases of research.

As an alternative, the measurement of *chlorinity was* proposed, based on the principle of the *Law of Constancy of Composition*, which states that for the main constituents of water, the proportions are virtually constant regardless of the absolute concentration of the total solids. This law makes it possible to find the salinity by simply determining the concentration of one of them. Chlorinity is defined as the amount in parts per thousand of chlorine, iodine and bromine expressed in grams per kilogram of seawater, assuming that the last two have been replaced by the first. Chlorinity is found by titrating sea water with a solution of silver nitrate in the presence of potassium chromate.

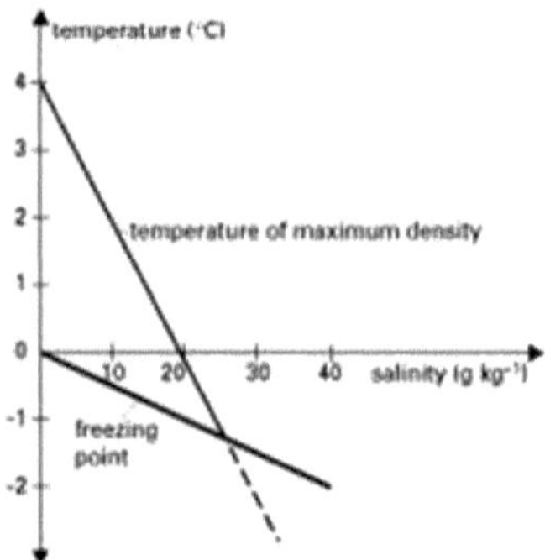

Figure 5.02. Graph depicting the decrease in maximum density temperature and freezing point with increasing salinity.

The definition of salinity was revisited again when techniques were developed to determine salinity from measurements of conductivity, temperature and pressure. It is known that the conductivity of seawater is strongly dependent on the salinity of the seawater. These techniques are much faster and can be applied in the field without the need for a chemical laboratory. UNESCO and the Institute

The *International Oceanographic Tables* giving the relationships between electrical conductivity, salinity and chlorinity have been published by the National Oceanographic Office of Great Britain. Since 1978, the Practical Salinity Scale has defined salinity in terms of a ratio or quotient of conductivities:

-The practical salinity, denoted by S, of a sample of sea water is defined in terms of the ratio, K, of the electrical conductivity of a sample of sea water at 15° and at the pressure of a standard atmosphere to that of a solution of potassium chloride (KCl), in which the total mass fraction of KCl is 0,0324356, at the same temperature and pressure. The value of K equal to one corresponds exactly, by definition, to a practical salinity equal to 35 l.

The corresponding formula is:

$$S = 0,0080 - 0,1692\ K^{1/2} + 25,3853\ K + 14,0941\ K^{3/2} - 7,0261\ K^2 + 2,7081\ K^{5/2}$$

In this definition, salinity is a quotient, so it has no unit, but the old value of 35^ corresponds to the value of 35 in practical salinity. Some oceanographers are still not used to using unitless numbers, so they write -35 psul, i.e. -practical salinity unitl, although this does not make any sense.

Whichever method is chosen for the determination of salinity, whether by titration (Strickland and Parsons, 1968), by conduction or by induction (UNESCO, 1971) or by the use of the refraction index, it is necessary to make this determination to three decimal places of accuracy. The accuracy of a single salinity determination by titration is ± 0,017%, by conduction is ± 0,005% and by induction is ± 0,003%.

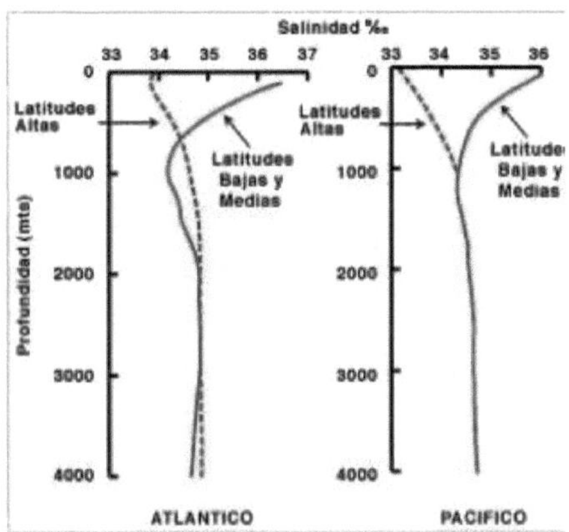

Figure 5.03. Salinity variation with depth.

Salinity in the oceans varies from 33% to 37%, with a global annual average of 34.7%, although it is often convenient to use the value of 35%. About 90% of the world's ocean volume has a much smaller range of 34% to 35% salinity. Deep water salinity varies within the 34.6% to 35% range (Figure 5.03). Salinity is higher at mid-latitudes while at low and high latitudes it is lower. The main processes responsible for this distribution are evaporation and precipitation at low and mid latitudes, so when evaporation exceeds precipitation, salinity is higher (as in the Red Sea, Mediterranean, Caribbean) and in areas of high precipitation it is lower, as at the equator.

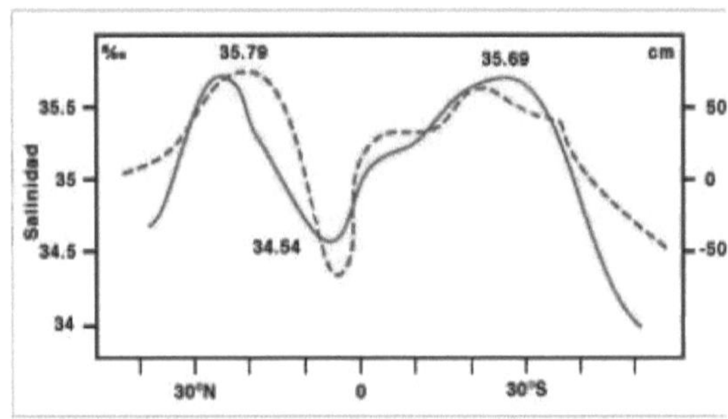

Figure 5.4. Variation of global mean surface salinity and the difference between evaporation minus precipitation with respect to latitude (Sverdrut et al., 1941).

This balance, together with the mixing process, gives a linear correlation with surface salinity (Fig. 5.4). The surface salinity curve for all oceans follows the evaporation curve, showing a maximum at 25°N of 35.79%, a minimum of 34.54% at 25°N and a minimum of 34.54% at 25°N.

5°N, a secondary maximum at 20 - 25°S of 35.69% decreasing sharply towards the poles. On the other hand, the freezing and melting of ice plays an important role in the distribution of salinity in the polar zones. Thus, where the ice melts, salinity decreases.

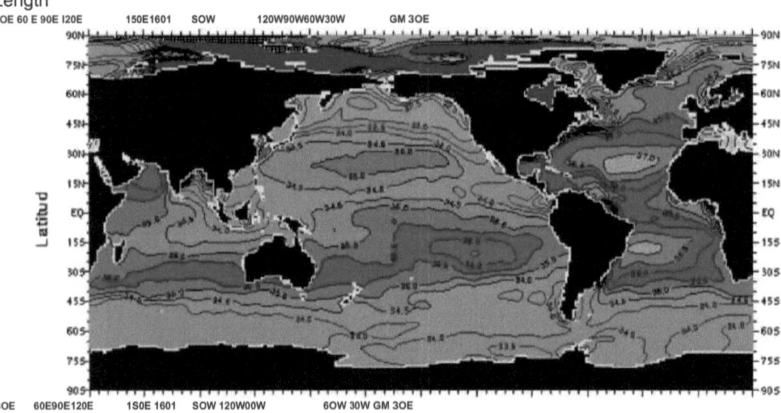

Figure 5.5. Mean annual sea surface salinity (ppm).

The variations of surface salinity, referring to large areas, depend mainly on variations in the difference between evaporation and precipitation considering annual periods. Diurnal variations have not been reported and are minimal, which in practice are neglected.

The distribution of salinity is zonal, but sea currents can introduce alterations to this pattern as can be seen in the salinity distribution graph presented in Figure 5.5.

3.4. Pressure

The mean pressure on any surface is defined as the force per unit area acting perpendicular to that surface. Pressure is a measure of the force acting within the ocean. Under the influence of gravity, pressure varies as a function of depth, and will be greater at the bottom of an ocean than at its surface. The unit

The most commonly used unit in oceanography is the decibar, although other units such as the pascal, atmosphere, etc. can be used. The highest pressures are found at depth because they support the greatest weight of water (hydrostatic pressure):

$$p = \rho\,g\,z$$

The pressure exerted per square centimetre in one metre of seawater is very close to 1 decibar, i.e. the hydrostatic pressure increases by one decibar with each metre of depth. Therefore, the depth in metres and the pressure in decibars have approximately the same value. This rule is sufficiently accurate to determine the effect of pressure on the physical properties of water, but for details of the pressure distribution it must be calculated from the density distribution.

For oceanographic purposes, the atmospheric pressure is always ignored, so the pressure at the sea surface is taken as zero. Pressure is essentially a function of depth and the numëeric value in decibars. The range in pressure will be from zero at the surface to about 11,000 decibars at the deepest part of the ocean.

Pressure has its origin at the molecular level, so it is considered a state variable. The line of equal pressure is known as isobars.

CHAPTER VI

FISKAS PROPERTIES

The impact of the océano on the global climate of our planet is beyond doubt. Many of man's activities are directly or indirectly linked to the ocean, as it is used as a means of transport, as a source of food, as a source of material and energy resources, etc.; hence man's interest in knowing and understanding it since the dawn of his presence on the planet. Its dynamics, as part of the earth's ecosystem, allows it to interact in a complex way, above all with the atmosphere, largely shaping the world's climate as we know it. Small and large-scale atmospheric events and phenomena occur continuously, some with a global impact, such as the phenomenon of the "Earth's climate". To understand how these phenomena occur it is necessary to understand their most elementary properties, how they are distributed both spatially and temporally.

The properties that most influence the oceans are the physical properties such as density, heat transfer, currents, etc., which are indicators that represent quite well the state of the sea at any given time. The physical properties in turn depend on the variables temperature, salinity and pressure which were discussed in the previous chapter.

Ocean is composed mainly of pure water, which accounts for 96.5%. The other 3.5% is made up of suspended particles, gases and dissolved salts of almost all the elements known on Earth. It is convenient to consider the structure of the water molecule as well as the chemical basis for the very peculiar properties that make it unique to this molecule, a necessary step to understand the roles that water plays in the physical properties of seawater.

We will therefore review some of the properties of pure water.

6.1. Properties of pure water.

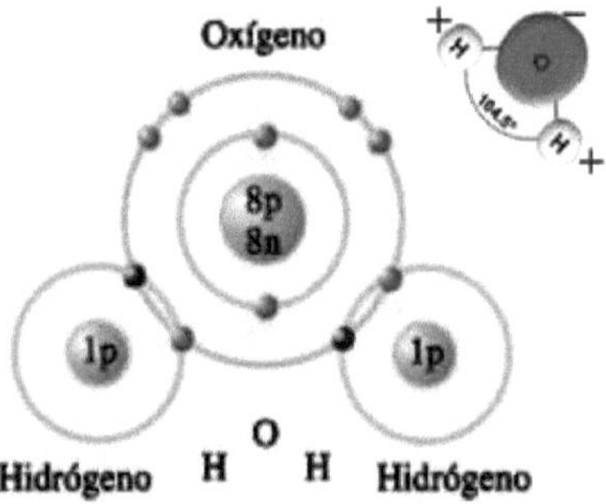

Figure 6.1. The water molecule. Source: http://www.fisicanet.com.ar/biologia/introduccion
_biologia/ap01_introduccion_a_la_biologia.php

The water molecule is very simple: it is composed of two hydrogen atoms and one oxygen atom. The bond between the oxygen and the hydrogen is of the covalent type.

Indeed, each hydrogen atom shares its single electron with the oxygen atom, while the oxygen atom attracts the two electrons it needs to complete its outer shell, thus creating a stable molecule. The two hydrogen atoms are bonded to the oxygen atom asymmetrically, forming an angle of 104.5° to each other. The immediate effect of this distribution is that the negative charge is concentrated on the oxygen (the electrons stay around its nucleus longer) and the positive charge on the hydrogen (the electrons stay around its nucleus for a shorter time).

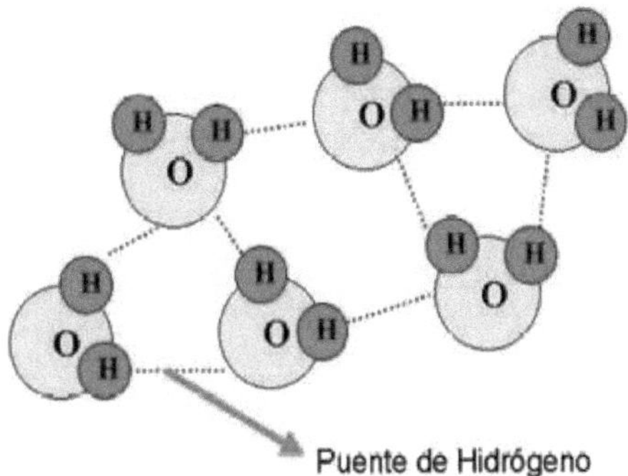

Figure 6.2. The water molecule. Source:
http://preupsubiologia.googlepages.com/2preupsubiologia

This polar nature of this molecule makes it behave like an ion. The hydrogen end (positively charged) attracts the oxygen side (negatively charged) of another adjacent water molecule. In this way *hydrogen bonds or hydrogen bridges* are formed between the adjacent molecules (fig. 6.2). These bonds are weak compared to the electron sharing bond (6%) so they are easily broken and restructured.

The hydrogen bonds and polarity of water molecules are responsible for many of the unique characteristics and physical properties of water:

If water were not polar, it would be a gas at room temperature and would have an extremely low freezing point (see figure 6.1). Thus,

> impossible to live in.

> At the air-water interface, the polar attractive nature of water causes the molecules at the surface to be strongly strsidssed by those below, allowing a film to form on the surface of the water, strong enough to support small objects. This phenomenon is known as surface tension and water has the highest surface tension of all common liquids, second only to mercury.

> The angle between the two hydrogen atoms is 104.6° (~105°), close to the angle of a tetrahedron, i.e. a structure with four arms emanating from a centre at equal angles (109°28'). The hydrogen bond between water molecules with a tetrahedral structure is strong enough to allow the oxygen atoms to have four hydrogen atoms associated with them in a tetrahedral arrangement. The electrostatic force between the positive and negative charge is large. Water molecules tend to interact strongly with each other, as well as with other ions. Hydrogen bonds need a bond energy 10 to 100 times smaller than molecular bonds, so water is very flexible in its reactions to change the chemical conditions.

> Due to hydrogen bonding, liquid water is a mixture of individual water molecules and clusters of water molecule bonds. They usually form aggregates of two, four and eight molecules. In contrast to the aggregation of other molecules which are more compact, tetrahedrons have by nature more open lattices. At high temperatures, single molecules or aggregates of two dominate. As the temperature decreases, larger aggregations increase, and these take up more space than those of the same number of molecules with smaller aggregations. As a result, the density of water peaks at 4°C.

> Water has a high capacity to hold heat energy, with the highest heat of vaporisation of most common substances (thus having a high boiling point allowing it to be taken up at the

Earth's surface relatively warm). When water evaporates, it absorbs considerable amounts of heat.

> Water has a high latent heat of fusion. When ice forms, a large amount of heat energy is released. Water therefore acts as a buffer against temperature changes and prevents the earth's climate from fluctuating rapidly.

> It has an unusually strong dissociation power, i.e. it separates the dissolved material into electrically charged ions. The charged ion is surrounded by dipole water molecules, known as hydration.

> The conductivity of pure water is relatively low. When it contains dissolved salts such as seawater, it increases significantly, being equivalent to half that of copper.

Solvent property of water. Water is the universal solvent. (How does the water molecule act? Due to its polar nature, the hydrogens of the water molecule form bonds between its hydrogen and sodium ions. In the case of seawater, if we consider a compound containing ionic bonds, such as sodium chloride, when that substance is placed in the water, the electrostatic attraction between the sodium and chloride ions is reduced by about 80 times. When more sodium and chlorine ions are released, due to the weakening of the electrostatic attraction that holds them together, they become surrounded by the polar water molecules. In this way the chlorine and sodium are separated by the water and dissolved in it.

Some thermal properties of pure water are:

Freezing and boiling point of water: The freezing point of water, i.e., when it changes from a liquid to a solid state, occurs at a temperature of 0°C. The boiling or vaporisation point of water is 100°C, where it changes from a liquid to a vapour. The freezing and boiling points of water are much higher than other similar molecules such as H_2Te, H_2Se and H_2S due to its dipole and hydrogen bonding.

Heat capacity: A direct result of hydrogen bonding is the high heat capacity of water. One calorie is the amount of heat required to raise the temperature of 1 g of water by 1°C. The heat capacity of water is large compared to that of most other substances.

Latent heat of liquefaction and vaporization: The unusually high heat capacity of water is closely related to its high latent heat of liquefaction and latent heat of vaporization. At the temperature where the change of state of any substance occurs there is no temperature rise even though heat is continually being added, as it is used entirely to break all the bonds necessary to complete the change of state. The amount of such heat required for 1 g of a substance, at the liquefaction point, to change from the solid to the rich state is called the *latent heat of liquefaction*. The heat applied to effect a change of state at the boiling point is the *latent heat of vaporisation*. The latent heat of liquefaction to convert 1 g of ice to 1 g of water is 80 Cal, and is higher for water than for any other substance. The latent heat of vaporisation required to convert water to steam is 540 Cal.

Surface tension: In a tube with a very small diameter, the cohesive force, which holds the water molecules together, in combination with the adhesive attraction force between the water molecules and the glass container, will pull the water column to great heights. This phenomenon is known as capillarity.

Table 6.1. Water temperature value at which some physical properties reach a minimum

Properties	T°C
Oxygen solubility	80
Specific volume	4
Specific heat	34

Properties	T°C
Solubility of hydrogen	37
Compressibility	44
Light speed	-1
Sound speed	74

The physical properties of most substances show a uniform variation with temperature. However, most of the physical properties of pure water show a maximum at a certain intermediate temperature. The speed of sound shows a maximum at 74°C.

On freezing, all water molecules form tetrahedrons. This leads to a sudden extension in volume, i.e. a decrease in density. The solid phase of water is therefore lighter than the liquid phase, which is a rare property. Some important consequences are:

> Ice floats. This is important for life in freshwater lakes, since ice acts as an insulator against additional heat loss, preventing the water from freezing from the surface to the bottom.

> Density shows a rapid decrease as it approaches the freezing point. The expansion that results during freezing is an important cause of rock weathering due to atmospheric action.

> The freezing point decreases with pressure. Hence, melting takes place at the base of glaciers, which facilitates glacier flow.

> Hydrogen bonds yield under pressure, i.e. ice under pressure becomes plastic. As a result, the ice that forms on land in the Antarctic and Arctic regions flows into the sea and forms icebergs at the outer edges. Without this process all the water in the world would eventually end up as ice in the polar regions.

6.2. Density

Density is defined as the ratio of the weight to the volume of a substance. The specific gravity is defined as the ratio of the density to the density of distilled water at 4°C. In oceanography, the term density p is generally used, although strictly speaking, it is always considered to be the specific gravity.

Density is a fundamental property for the study of ocean dynamics. Small horizontal differences cause currents. The density of seawater depends on three variables: temperature, salinity and pressure ($p_{t,s,p}$). Density increases with increasing salinity and decreasing temperature (except at temperatures below the density maximum). In 1902 Knudsen and Ekman developed the equation of state expressing density in units of g/cm^3, as a function of its variables. In 1980, the international equation of state was established.

$$\rho = f(T, S, p)$$

Where:

T = temperature, in °C

S = salinity, psu

p = pressure, in decibars

ρ = density, kg/m^3

A density of 1.025 g/cm^3 in the 1902 formula corresponds to 1025 kg/m^3 in the 1980 formula. The average density of seawater is 1025 kg/m^3.

In physical oceanography it is necessary to measure density to at least five decimal places. In order to minimise the errors in the recording and for convenience, the *sigma* system is used instead of the density. In this system, in *situ* density $^{(\rho_{s,t,p})}$ is replaced by *in situ sigma* $^{(\sigma_{s,t,p})}$.

. The sub-indices s, t and p indicate the salinity, temperature and depth (or pressure), respectively, of the sample. The *in situ* sigma is a measure of the density of the sample at the sampling site and is defined by

$$\sigma_{s,t,p} \left(= \rho_{s,t,p} - 1 \right) \times 1000$$

If the density is referred to the sea surface where the hydrostatic pressure is zero, it is denoted by *sigma-t*[4] which has the following definition:

$$\sigma_t = \left(\rho - 1000 \right) \mathrm{kg \cdot m^{-3}}$$

for temperatures from -2 to 30°C and salinities from 20^ to 40^ covering the ranges found in open oceans. This is mainly subsurface water, the rest of the range representing limited volumes of surface water.

At the surface, density is determined by temperature and salinity: it is proportional to salinity and inversely proportional to temperature; this relationship is not linear, more so for temperature than for salinity. The saltiest and coldest waters are the densest, hence the deepest. As salinity varies less than temperature in the open ocean, the density of water at the surface depends more on the latter.

On the other hand density is less sensitive to temperature changes at low temperatures than at high temperatures. Note that pure water has a maximum density around 4°C at atmospheric pressure, but for seawater it decreases due to salinity. It would be obtained at -3.7°C for water with a salinity of 35^ below the thawing point. This explains why, in the sea, the density of water increases continuously as the temperature decreases and why surface waters freeze before reaching the density threshold that would make them denser than bottom waters, which prevents homothermia by vertical mixing. A useful rule of thumb is that density increases by about 1 part in 1000 (i.e. by 1 kgjn^{-3}) for a temperature change of -5°C, for a salinity change of +1 or for a pressure change of +200 dbar (equivalent to a depth increase of about 200 m).

In the vertical direction, the density varies not only with temperature and salinity, but also increases with pressure. It increases rapidly with depth. A column of water of height 1 m and surface area 1 dm^2 exerts a pressure of 10.3 k (0.103 k/cm^2). At 1000 m it will be 103 k/cm^2 . In summary, density increases with depth as temperature decreases, but more so in low latitudes where the difference between surface and surface is greater (24 to 27.9) than in cold zones where the gradient is weaker (27 to 27.9). This has two very important hydrological consequences:

> The vertical movements of water generated by density differences are facilitated more in the temperate and cold regions than in the warm zone where the high temperatures of the surface layers retain water of low density that cannot sink to the depths.

> The high-density waters that fill the bottom of the oceans are formed at the surface, generating a predominantly meridional circulation, while that of the surface waters due to the influence of the winds is mainly latitudinal.

Since waters of different densities do not mix, or mix only very slowly, it is customary to distinguish between water masses according to their temperature and salinity, i.e. on the basis of their density. For this purpose, temperature and salinity profiles are established according

---

[4] This term was introduced as an abbreviation. The density of seawater at atmospheric pressure varies from about 1000 kg-m^3 for near fresh water to about 1028 kg-m^3 for the densest oceanic waters. As the variation is in the last two digits only for descriptive purposes only these two digits are used.

to depth, the so-called TS diagrams, which associate temperatures as ordinates and salinities as abscissae on the same graph. The

The existence of water masses with different physical properties is, together with the winds, the driving force of the general ocean circulation. These water masses acquire most of their properties at the surface, in contact with the atmosphere.

In situ density can be determined by the expressions of Knudsen (1901), Forch et al. (1902) and Ekman (1908) using determinations of sampling depth (pressure), salinity and sample temperature. Knudsen (1901) determined the specific gravity anomaly.

In many oceanographic publications the term density is sometimes used (ρ), sometimes relative density (d), sometimes specific volume ($\alpha = 1/\rho$) and sometimes -sigma-t). But it should be kept in mind that the quantity obtained from polynomial expressions or tables using temperature, salinity and pressure values is, strictly speaking, the relative density because they are based on comparisons of seawater with pure water, not absolute density measurements which are much more difficult to measure. Relative density can be given with an accuracy of about 3 in 10^6 but proper density is only accurate to 10 in 106. Fortunately in most cases it is the density differences that are important.

The *specific volume* (α), as mentioned above, is the reciprocal of the density $(\alpha = 1/\rho)$ and has the units of m^3 kg^{-1}. In oceanography, two other densities of related quantities are used, *specific volume anomaly* (δ) and *thermosteric anomaly* $[\Delta_{s,t}]$ which will be defined later.

Density and specific volume as a function of temperature, salinity and pressure. For a single component fluid, the density determination is $\rho = \rho_{(T,p)}$, where T is the absolute (thermodynamic) temperature and the specific volume $\alpha = \alpha_{(T,p)}$, i.e. they are a function of temperature and pressure only. For seawater, which is a multicomponent fluid, the dissolved salts add further complication and $\alpha = \alpha_{(s,T,p)}$. For seawater, it is usual to use the Celsius temperature *T*, rather than the absolute temperature T, and the relation $\alpha = \alpha_{(s,T,p)}$ can be represented as a table with the observed values of *S*, *T* and *p* or as a polynomial in these parameters.

From the first measurements of density variation with temperature, salinity and pressure it was found that the most convenient way to express the results was in terms of specific volume, as follows:

$$\alpha_{(s,t,p)} = \alpha_{(35,0,p)} + \delta_s + \delta_t + \delta_{s,t} + \delta_{s,p} + \delta_{t,p} + \delta_{s,t,p}$$

$$\alpha_{(s,t,p)} - \alpha_{(35,0,p)} = \delta = +\Delta_{s,t} + \delta_{s,t} + \delta_{s,p} + \delta_{t,p} + \delta_{s,t,p}$$

In these expressions, $\alpha_{(s,t,p)}$ is the specific volume of a water sample of salinity *S*, temperature *T* and sea pressure *p*. Then $\alpha_{(35,0,p)}$ is the specific volume of an arbitrary standard seawater of *S=35*, *T=0°C* and pressure *p* at the depth of the sample. This term expresses most of the effects of pressure on the specific volume. Term 5, anomafia of the specific volume, represents the sum of the six anomafia terms in the first equation. The quantity $\Delta_{s,t} = \delta_s + \delta_t + \delta_{s,t}$ accounts for most of the salinity and temperature effects, regardless of pressure, and is called *thermosteric anomafia*. The terms $\delta_{s,p}$ y $\delta_{t,p}$ account respectively for most of the combined effects of salinity and pressure and temperature and pressure. The latter term $\delta_{s,t,p}$ is so small that it is always neglected. For waters shallower than 1000 m depth, the thermo-thermal anomorphism, $\Delta_{s,t,}$ is the largest component of δ and the pressure terms $\delta_{s,p}$ y $\delta_{t,p}$' can often be neglected. In recent years $\Delta_{s,t}$] has to some extent replaced $\sigma_{t,}$, as a parameter for describing density characteristics in the upper octian layer because it can be used more directly than σ_t in first-order dynamics calculations.

From $\alpha(s,t,0)$ ó $1/\rho(s,t,0) = 1/(1000 + \sigma_t)$ y $\alpha(s,t,0) = \alpha(35,0,0) + \Delta_{s,t}$: and observing that $\alpha(35,0,0) = 0{,}97266 \times 10^{-3}\ \mathrm{m^3 kg^{-1}}$ it is easy to demonstrate that:

$$\Delta_{s,t} = \left(\frac{1000}{1000 + \sigma_t} - 0{,}97266*10^{-3} \right)\ m^3\ kg^{-1}$$

Some values are as shown below:

$\sigma_t =$	23,00	24,00	25,00	26,00	27,00	28,00	kg m^{-3}
$\Delta_{s,t}=$	485,7	390,3	295,0	199,9	105,0	10,3x10^{-8}	m^{3}kg^{-1}

Density variation, even in small proportions, can produce currents. Thus, vertical movement can bring cold water from deeper subsurface levels, or can cause surface water to submerge. At different locations, different processes exert a controlling influence on the annual temperature picture, so the large-scale picture is complex.

The salinity of water in any region depends mainly on the relative amount of evaporation and precipitation. Evaporation makes the water pure, thus increasing the salt concentration in the surface layers. Turbulent mixing distributes the salinity changes downwards. Precipitation dilutes seawater, and in some regions where river discharges are important, they are also important in reducing the salinity of the sea. Currents play an important role in carrying high or low salinity water to other regions. In winter and at high latitudes there is another process that affects salinity. Sea ice is fresh, and therefore its cooling renews as much pure water as evaporation does, increasing the salinity of unfrozen water. When the ice melts, salinity is reduced by dilution.

The freezing process is interesting. When the temperature decreases to the freezing point, pure ice crystals begin to form and coalesce into a matrix. Once the freezing process starts, it continues rapidly due to the radiation deficit and some pockets of salt water are mechanically trapped inside the thickened ice. Salt water freezes at a lower temperature than fresh water and as the temperature drops the pockets of water gradually shrink, concentrating the brine, until finally the salt can crystallise, however, these pockets of brine are the weakest points of ice. At any rise in temperature they are the first to melt. Moreover, because water expands when it freezes, it will melt simply by compression. The pile-up

The ice formed by the wind causes melting of the weakest points, which are the brine pockets, causing them to drain out of the block. Thus, except for newly formed ice that has not been piled up by the wind or exposed to temperature changes, seawater ice is fresh.

The temperature and salinity of the water determine its density and the density distribution is closely related to the circulation of the oceans. In a vertical direction, density must increase with depth, or else overturning and readjustment will take place in the water. In a horizontal direction, density variations can be maintained only if the water moves horizontally. Conversely, wherever a permanent stream is flowing, the density must vary from one part of the stream to the other.

3.6. Thermal properties of seawater.

They are:

Thermal expansion. The coefficient of thermal expansion defined by:

$$e = \left(\frac{1}{\alpha_{s,t,p}} \right)\left(\frac{\partial \alpha_{s,t,p}}{\partial t} \right)$$

is obtained at atmospheric pressure from the terms for D in Knudsen's hydrographic tables (Sverdrup. 1941). The coefficient for seawater is higher than for pure water and increases with increasing pressure.

Thermal conductivity. In water the temperature varies with space. Heat is conducted from regions of higher temperature to regions of lower temperature. The amount of heat in calories grams per second that passes through a surface area of cm^2 is proportional to the change in temperature per centimetre along a line normal to the surface. The coefficient of proportionality, y, is called the coefficient of thermal conductivity ($dQ/dt = -y\ dt/dn$). For pure water at 15°C the coefficient is equal to 1.39×10^{-3} . The coefficient is somewhat smaller for seawater and increases with increasing temperature.

increase in temperature and pressure. This coefficient is valid, however, only if the water is at rest or in laminar motion, but in oceans the water is almost always in a state of turbulent motion in which the heat transfer processes are completely altered. In these circumstances the coefficient of heat conductivity must be replaced by a coefficient of turbulence which is several times higher and depends mainly on the state of motion than on the effects of temperature and pressure.

Specific heat. Specific heat is the number of calories required to increase the temperature of 1 gram of a substance by 1°C. When studying fluids. The specific heat at constant pressure, cp, is usually a measurable property, although in certain problems the specific heat at constant volume, cv, must be known. An empirical equation for the specific heat at 0°C and atmospheric pressure:

$$C_p = 1,005 - 0,004136\ S + 0,0001098\ S^2 - 0,000001324\ S^3$$

It is observed that the sediment heat decreases with increasing salinity, although the effect is somewhat larger than would be expected from the composition of the solution. The effect of temperature on the spedaphic heat does not alter it significantly. The effect of pressure on the sediment heat has been calculated from the following formula:

$$\frac{dc_p}{dp} = -10^5\ \frac{T}{\rho}\ \frac{1}{J}\left(\frac{de}{dt} + e^2 \right)$$

where p is the pressure in decibars, T is the absolute temperature, p the density, J the mechanical equivalence of the heat and - the coefficient of thermal expansion. The specific heat at constant volume, which is somewhat less than c_p, can be calculated from the following equation:

$$c_t = c_p - \frac{Te^2}{\rho KJ}$$

where K is the true compressibility. The ratio of $c_p{:}c_v$ for water of 34.85^ at atmospheric pressure increases from 1.0004 at 0°C to 1.0207 at 30°C. The effect of pressure is appreciable; for the same water at 0°C, the ratio is 1.0009 at 1000 decibars to 1.0126 at 10,000 decibars. This ratio is important in the study of the speed of sound.

Latent heat of evaporation. The latent heat of evaporation of pure water is defined as the amount of heat in calorie grams needed to evaporate 1 g of water, or as the amount of heat needed to produce 1 g of water vapour at the same water temperature. Only the latter definition is applicable to seawater. The latent heat of evaporation of seawater does not differ much from pure water, therefore, between the temperatures of 0°C and 30°C the formula can be used:

$$L = 596 - 0{,}52\,t$$

Adiabatic temperature change. When a fluid is compressed without heat gain or loss, work is done in the system causing a rise in temperature. If expansion occurs, the fluid itself gives off more energy, which is reflected by a drop in temperature. This adiabatic temperature change is widely known and is important in the atmosphere. Seawater is compressible and the effects of the adiabatic process, though small, must be taken into account when studying the vertical temperature distribution in the deep ocean and any adiabatic effect is related to changes in pressure. But in the sea the pressure can be considered proportional to depth and the adiabatic temperature changes are given as changes per unit depth rather than per unit pressure. According to Lord Kelvin, the change in temperature per centimetre of displacement is:

$$10^{-5}\,dt = \frac{Te}{Jc_p}\,g\rho$$

where T is the absolute temperature and g is the acceleration of gravity and where the other symbols have already been defined. This change is extremely small and for practical purposes the adiabatic temperature change over a vertical distance of 1000 m, called *the adiabatic temperature gradient,* is used. It should be noted that the adiabatic temperature gradient depends mainly on the coefficient of thermal expansion tërmica, varying much more with temperature and pressure than with other quantities involved.

The temperature that a water sample would obtain if brought adiabatically to the sea surface is known as the *potential temperature* and is designated as 0. Thus $O{=}t_m{-}\Delta t$, where t_m is the *in situ* temperature and Δt is the amount by which the temperature would decrease adiabatically if the sample were raised to the surface. The potential temperature can be obtained from the table of adiabatic gradients by tedious calculations, although tables have been prepared to find Δt. This table is based on a salinity of 34.85‰ ($\sigma_0{=}28{,}0$) and is generally applicable to deep ocean regions, because in these the salinity does not differ much from 34.85‰ and because the effect of salinity on adiabatic processes is small. It could be that if 2°C water is brought adiabatically from 8000 m to the surface, $\Delta t=0.925°$ and thus the potential water temperature is 1,075°.

6.4. Electrical conductivity

The conductivity of seawater depends on the number of dissolved ions per unit volume (i.e. salinity) and the mobility of the ions (i.e. temperature and pressure). Its units are mS/cm (milli-Siemens per centimetre). The higher the amount of dissolved salts, the higher the conductivity, this effect continues until the solution is so full of ions that the freedom of movement is restricted and the conductivity may decrease rather than increase, with two different concentrations having the same conductivity. Conductivity increases by the same amount with an increase in salinity of 0.01, an increase in temperature of 0.01°C and an increase in depth (i.e. pressure) of 20 metres. In most practical oceanographic applications the change in conductivity is dominated by temperature.

Table 6.2. Conductivity values of some typical samples

Sample temperature at 25 ° C	Conductivity, pS/cm
Ultrapure water	0.05
Boiler feed water	1 a 5
Drinking water	50 a 100
Seawater	53,000
5 % NaOH	223,000
50 % NaOH	150,000
10 % HCl	700,000
32 % HCl	700,000
31 % HNO_3	865,000

Some substances ionise more completely than others and therefore conduct current better. Each acid, base or salt has its own characteristic curve of concentration versus conductivity. Good conductors are: acids, bases and inorganic salts: HCl, NaOH, NaCl, Na_2CO_3, etc. Bad conductors are: molecules of organic substances, which by the nature of their bonds are non-ionic and therefore do not conduct electric current. An increase in temperature decreases the viscosity of water and allows ions to move faster, conducting more electricity. This temperature effect is different for each ion. Knowing these factors, measuring conductivity allows us to get a very good idea of the amount of dissolved salts.

6.5. Acoustic properties

Sound propagates along rays (in the same way as light does). Thus, the laws of geometrical optics apply equally to sound.

1.   Sound travels along straight trajectories where the speed of sound c is constant; otherwise the trajectory deviates towards the region of lower c values.

2.   Different beams are independent of each other.

3.   Sonic trajectories are reversible.

4.   The law of reflection (angle of incidence = angle of reflection) holds true at the bottom of the sea, on the sea surface, on objects and surfaces.

$$\frac{\sin \alpha_1}{\sin \alpha_2} = \frac{c_1}{c_2}$$

5.   The law of refraction is fulfilled at interfaces:

Since the stratification in the ocean is almost horizontal, the propagation of sound in the vertical is practically along a straight path. This is the basis of echo sounding: The depth can be known as long as the average sound speed is known. A first estimate is 1500 m s^{-1} ; tables exist which provide corrections of c for different areas of the world's oceans.

Its velocity, c, depends on temperature, salinity and pressure and varies between 1400 ms^{-1} and 1600 ms^{-1} . In the open ocean c is influenced by the distribution of temperature and pressure, but not much by salinity. It decreases with decreasing temperature, pressure and salinity. The combination of the variation of these three parameters with depth produces a vertical profile of the speed of sound: Temperature decreases rapidly in the upper kilometre of the ocean affecting the speed of sound, i.e. c decreases with depth. In the deeper regions (below the upper kilometre) the temperature change with depth is small and c is determined by the increase in pressure, i.e. c increases with depth. Vertical changes in salinity are too small to have an impact; but the mean salinity is determined if c has a low value (if the mean salinity is low) or a high value (if the mean salinity is high) on average.

Figure 6.3 gives examples of horizontal sound paths. The first diagram shows the propagation of sound at the depth of the minimum sound speed (around 1000 m depth). The sound rays are deflected back towards the depth of the -minimum sound speed and travel long distances at that depth (they can pass through all the oceans). The sonic channel is known as the SOFAR (SOund Fixing And Ranging) channel. Before the introduction of the Global Positioning System (GPS), the SOFAR channel was used to locate ships and aircraft in distress situations, and to track floats in the study of ocean currents. The second diagram presents a situation where the homothermal mixing layer (about 100 m thick) is above the normal temperature stratification. In this case the speed of sound increases due to the increase in pressure before it decreases due to the decrease in temperature. The maximum speed of sound (around 100 m depth) creates a shadow zone, because all sonic rays are deflected away from that depth. Sound is an information carrier used in human and animal communication. Sound can travel long distances and is therefore used for various purposes, such as deep sounding, communication, searching for objects and underwater measurements, by animals and by man.

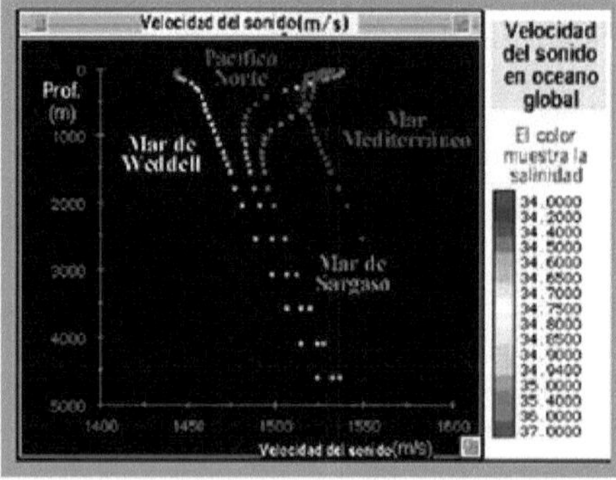

Figure 6.3. Speed of sound as a function of depth in various oceanic regions. Near the surface, c decreases as a result of decreasing temperature with depth. At greater depth the temperature changes are smaller, so c increases linearly with pressure (depth). Although the effect of salinity on the variation of c with depth is insignificant, salinity nevertheless determines the overall magnitude of c: Salinity is low in the Pacific Ocean, higher in the Sargasso Sea (Atlantic Ocean) and highest in the Mediterranean Sea. The Weddell Sea is an example of a polar region: The temperature is uniform over the entire depth range and is extremely low, so that c is also low and shows only a vertical dependence on pressure (depth).

3.9. Compressibility.

The capacity of water to undergo compression by altering its volume. Since the beginning of the 20th century, mathematical expressions have been formulated to describe the average compressibility of seawater, such as the one given by Ekman in 1908:

$$\alpha_{s,t,p} = \alpha_{s,t,0}\left(1 - kp\right)$$

The true compressibility of seawater is described by a coefficient representing the proportional change in specific volume if the hydrostatic pressure is increased by one unit of pressure. It is calculated using the following equation:

$$K = \frac{\left(k + p\dfrac{dk}{dp}\right)}{\left(1 - kp\right)}$$

where k is the average compressibility when bar is used as the unit of pressure and p is the pressure in bar.

CHAPTER VII

SURFACE LAYER AND WINDS

Of all the oceanic mass, the surface layer is the most active, in almost every sense. Indeed, it is in this layer that the largest ocean currents occur, it is home to the highest percentage of living organisms, and it has the greatest impact on global climate and human activity.

What do we mean by surface layer? For the purpose of the present work, it corresponds to the first layer of the sea, which is widely divided, ranging from zero to two hundred metres deep, and which is exposed to the direct influence of the atmosphere.

What do we mean by wind? It is the displacement of air masses due to a difference in atmospheric pressure between two points. The origin of the winds is found in the difference in temperature due to the unequal heating of the atmosphere by the sun, the hot air tends to rise while the cold air tends to sink.

We will first try to describe the characteristics and properties of both the sea surface layer and the wind. Then we will briefly outline the types of interactions that occur between them and the processes involved. In particular, surface currents and sea waves will be discussed. Finally, the effects of this interaction on marine flora and fauna, as well as on fisheries, will be described.

7.1. The surface layer, winds and their interactions

The surface layer. As shown in the attached image, it corresponds to the layer that is in contact with the atmosphere. For this reason, it is the most dynamic of the rest of the ocean. It is in this layer that the main ocean currents and waves, evaporation and precipitation, heating of the water by the sun, gas exchange by diffusion or mixing processes occur. All this leads to the water masses acquiring their own characteristics, which in turn determines the presence and distribution of plant and animal life in the oceans. It is in this layer that man exerts his influence and makes use of it, whether to navigate, to obtain food or to extract minerals or energy to sustain human civilisation.

In this chapter we will only focus on the interaction with the wind, which is responsible for the generation of currents and waves as well as the influence of these interactions on living beings.

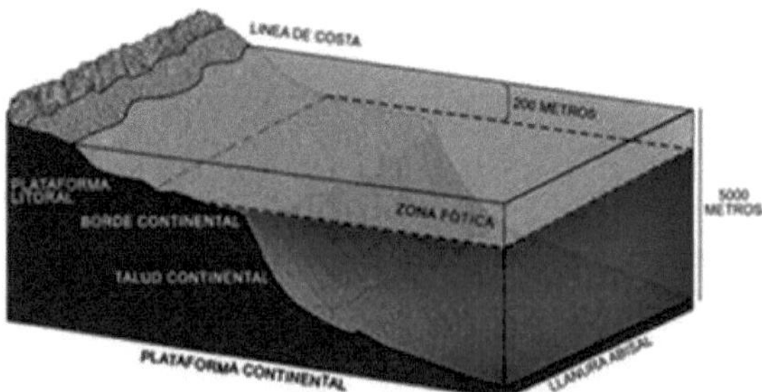

Figure 7.1. Surface layer of the oceans, typically 200 m deep. Source: http://socialesguada.weebly.com/definiciones-geografa.html

The wind. On atmospheric circulation, the general circulation on a large scale was described in chapter 3. At the local scale, at the surface level, the wind is defined by two parameters: the direction in the horizontal plane and the velocity. At the meso-scale, there is a strong relationship between the solar becomes energy стёйса of the winds. Figure 7.2 shows the average winds at the surface level, especially their action on the oceans.

Interaction between the surface layer and the winds. The surface of the oceans is in continuous contact with the atmosphere, which is often in motion in the form of winds. The action of the wind on the sea surface is constant and produces various effects.

When the wind is constant and sufficiently intense, such as the trade winds, it produces ocean currents. Local winds are mainly responsible for the waves. These winds tend to generate mixing processes in the surface layer, which when occurring in a water mass front zone facilitates high primary productivity. On the other hand, the combined action of the winds with the Coriolis force produces the Ekman Transport which in turn is responsible for the coastal upwelling that occurs on the eastern sides of the ocean basins.

Ocean surface currents are the main vectors of movement of the masses of the entire ocean, responsible for the transport of heat, distribution of temperature, salinity, oxygen and other chemical components on the surface of the oceans. Here we will not deal extensively with the subject of ocean surface currents because the next chapter will deal with it in detail.

7.2. The waves of the sea

Waves are generated by the wind. In chapter 3 it was already explained how waves are generated and that in reality there is no movement of water particles. The movement of the waves and the movement of the water particles that form them are different. The waves move over the surface at a speed determined by their length (or period) while the water particles move in vertical circles. Each circle represents the orbit of a single water particle on the surface of the sea and its diameter is equal to the height of the wave.

The water particles below the surface also move in circles, but the size of the water particles is not very large.

wind and ocean currents. According to some scientific studies, 2% of the energy

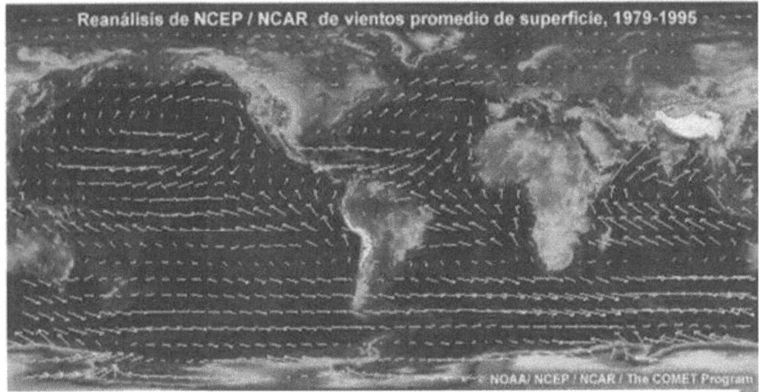

Figure 7.2. General circulation of the atmosphere. Source: NCEP / NCAR

is converted into kinetic energy of the winds. Figure 7.2 shows the average winds at surface level, especially their action over the oceans.

Interaction between the surface layer and winds. The surface of the oceans is in continuous contact with the atmosphere, which is often in motion in the form of winds. The action of the wind on the sea surface is constant and produces various effects.

When the wind is constant and sufficiently intense, such as the trade winds, it produces ocean currents. Local winds are mainly responsible for waves. These winds usually generate mixing processes in the surface layer, which when occurring in a water mass front zone facilitates high primary productivity. On the other hand, the combined action of the winds with the Coriolis force produces the Ekman Transport which in turn is responsible for the coastal upwelling that occurs on the eastern sides of the ocean basins.

Ocean surface currents are the main vectors for the movement of masses throughout the ocean, responsible for the transport of heat, temperature distribution, salinity, oxygen and other chemical constituents at the surface of the oceans. We will not deal extensively with ocean surface currents here because the next chapter will deal with them in detail.

7.2. The waves of the sea

Waves are generated by the wind. In chapter 3 it was already explained how waves are generated and that there is no actual movement of the water particles. The movement of waves and the movement of the water particles that form them are different. The waves move over the surface at a speed determined by their length (or period) while the water particles move in vertical circles. Each circle represents the orbit of a single water particle above the sea surface and its diameter is equal to the height of the wave.

Water particles below the surface also move in circles, but the size of their orbits decreases exponentially with depth. At a

depth equal to half the length of the wave, the size of the orbits is only 1/23 that of the surface. Below this depth, the motion is negligible.

Waves always advance faster than water particles, due to the difference in distance travelled per period. That is, if a wave is 5 metres high and has a wave length of 100 hundred metres, in one period this wave would advance 100 metres, while a water particle would advance only 5 times the period or about 16 metres (the distance around its orbit).

The longer and lower the waves are, the greater the difference in velocity between them and the water particles. In fact, the orbits of the water particles are not true circles. As the wave passes, the water particles move forward on the crest a little further than when they recede in the sinus, so they do not return to exactly the same point. There is a slow progression of the water in the direction of the wave movement.

It is greater in steep waves than in the long, low waves of the levelling sea. For a better understanding, we will give the definitions of some terms describing waves:

- **Wave height (H)**: The vertical distance between the sine and the crest of the wave (twice its amplitude).
- **Wave length (L)**: Horizontal distance between two successive crests.
- **Wave period (T)**: Time interval between two consecutive crests passing a fixed point. It is the easiest to measure.
- **Wave speed (C)**: Distance travelled per second.

On the other hand, we can establish two types of waves:

- **Wind waves**: These are generated by local winds and grow by the action of the waves. They are characterised by their short, steep, crested cusp, their irregularity and often break into white foam.
- **Bottom or cam waves**: These are waves that move out of the area in which they are generated, as they move they lose height and slowly decay, becoming smoother, more regular, with longer wave length and long crests. They can travel thousands of kilometres in the ocean and maintain their height and power.

When the wind stops acting on the waves, the wind waves stop growing and become cam waves. On the other hand, if high cam waves move before the wind starts to blow, the growth of the cam waves may be greater than if they started in a calm sea.

Development and growth of wind waves. The development of wind waves depends on wind speed and duration, as well as on the initial sea state and the extent of water over which the wind blows (fetch)*. These factors can be determined with meteorological maps of the ocean and the period and height of the resulting waves can be determined by means of graphical methods.

In the fetch, the wind picks up waves of many lengths, depending on the strength of its gusts. Each group of waves moves independently of any other group that may be present. The interference between waves causes them to move at different speeds, and often in different directions, making the sea appear rough and stormy when the wind is blowing very hard. As the wind continues to blow over the waves, they increase in height and speed (hence in length and period). The wind acts on the waves in two ways: by pushing them directly over the back side and by dragging the water particles tangentially forward.

Shorter waves, which move more slowly, receive the greatest direct push from the wind and therefore grow rapidly in height. However, there is a critical growth beyond which waves cannot grow. When the height is 1/7th of its length, the wave becomes unstable and breaks apart until it is destroyed. In stormy weather, waves rarely reach this critical steepness. When the wave reaches a height of about 1/10 of its length, the wind usually blows over its crest and diminishes the top of it. As the smaller waves rapidly steepen and break, the longer waves become dominant. Larger waves grow slowly, but reach greater heights before breaking. Longer waves move at about the same speed as the wind, and therefore receive very little thrust from ëste and remain low for a long time.

time. According to a simple rule of thumb, when the wind has been blowing for several hours, the prevailing waves have a period in seconds approximately equal to 1/4 of the wind speed gradient in knots, and they travel at about 3/4 of the wind speed.

The forward tangential drag of the wind on the water particles causes the waves to move faster than the wind itself. Even in this case the water particles move much slower than the wind. When the waves move for a longer period of time under the same wind, the speed of the waves will gradually increase to almost Ш times the wind speed. This can happen in the region of the trade winds, where they blow over huge expanses of the ocean, and also in the westerly wind regions of the southern hemisphere.

This is not the case in the northern oceans, where winds rarely blow long enough to allow full wave development. Waves exceed wind speeds more frequently during light winds, because strong winds are of shorter duration. The average wave height in all oceans is about 1.2 metres. Waves of 6 metres in height are common in the westerly wind zones of both hemispheres during the storm season. Waves higher than this are rare. For waves higher than 12 metres to exist, there would have to be a very strong wind blowing over an "extension" of 101500 kilometres.

Calculation of waves: In the sea, the period of the largest waves can be calculated by taking the time of successive rises of foam patches or floating objects. To obtain a reliable value, the average time of ten consecutive rises should be taken; then repeat this operation. The period can be easily obtained from the beach by taking the time between breakers, knowing that the period is the characteristic that does not change as long as the waves advance in shallow water. In deep water there is a close relationship between period, wave speed and wave length. If the period is expressed in seconds, the following equations will give a close approximation:

C (knots) = 3 T

C (ft/s) = 5 T

L (feet) = 5 T^2

When the waves are moving in shallow water, less than half the wave length, the above mentioned scale has no value, because the speed and length of the waves start to decrease while the period remains constant. Only the following equation would be valid in this case:

$$C = \frac{L}{T} \; pies/s$$

Sometimes certain rules of thumb are useful. One relates the period of the waves to the wind speed:

T (s) = 1/4 wind speed (knots).

Another, indicates that if the wind is strong, but the "extension" is limited.

H (feet) = 1 1/2 (nautical miles).

There are two empmca formulas that relate the height of the largest waves to the wind speed. In strong winds:

Hmax (feet) = 0.8 wind speed (knots).

In any wind:

Hmax (feet) = 0.026 (wind speed)^2 (knots).

Cam waves: As the waves leave the "extension" and spread out into areas where the wind is so light that it cannot sustain them, their period continues to increase, but their height slowly decreases. Cam waves become more

regular and gentle. There are two ways of classifying waves: One, according to their length. Since larger waves move faster, they tend to be ahead of the smaller waves. This contributes to their smoothness and regularity. Likewise, in cam waves approaching the coast, the ones with the longest period arrive first, followed by the ones with the shortest period, even though they all come from the same storm.

The other is by damping the waves according to their length. Waves lose height due to air resistance as they advance. Smaller waves lose their height (until they disappear) faster than larger ones. A rough rule of thumb is that waves lose 1/3 of their height over a distance in miles equal to their length in feet. This process contributes not only to the smoothness of the waves, but also to the gradual increase in length, period and speed of the cam wave train as a whole.

A wave train in deep water moves at about half the speed of an individual wave. This is observed when a stone is thrown into a pond. In a large wave train, the length, speed and period of ëstas are not constant, but increase rapidly towards the front of the train.

Not only the energy of the wave, but its speed and period advance with half the speed of the waves at those points. This requires that the individual waves advancing within the train show a slow and steady increase in length, speed and period. There are often several trains of waves moving over the sea surface at the same time, coming from different storms or from different parts of the same storm. Interference with each other causes groups of high waves and groups of low waves to form. If 2 cam wave trains move in the same direction and are somewhat different in length, the resulting wave length would be an average, but the heights would be additive. There will also be waves where the crests and troughs coincide, and the surface ups and downs equal the sum of the heights of two trains, causing large waves to develop. Between these lines the crests and troughs of two trains partially cancel each other, thus developing low waves, lower than for each train separately.

Another example is when a wind of short duration blows over a long and low levelled sea. Only the short action is apparent, but it will be of a slow and variable height. These types of high and low wave interference also move at half the speed of the individual waves.

Cam wave forecasting: Having good weather maps of the ocean, the height and period of the waves can be used to determine the time of the spread. The same maps will show the distance, from the extent to the coast, at which the cam wave will arrive and the wind conditions involved. The height, period and arrival time of the wave can be predicted if the changes in fetch as it travels are known. Although problems can arise from wind variability, a moving fetch or from winds between the fetch and the coast, there are very satisfactory cam sea forecasts. Their relationships are determined graphically.

The time taken to travel from the fetch to the shore is calculated by means of formulas. The speed of the cam train, equal to half the speed of the wave, is equal to the distance travelled divided by the time taken:

$$1/2\ C\ (nudos) = \frac{distancia\ (millas\ náuticas)}{Tiempo\ recorrido\ (horas)}$$

$$Tiempo\ (horas) = \frac{distancia}{1/2\ C} = \frac{2}{3}\frac{distancia\ (m.n.)}{periodo\ (s)}$$

Example:

Assuming that the storm occurs in the Gulf of Alaska (about 2000 nautical miles out) and the period of the incoming sea level is 13 seconds, we have:

$$Tiempo = \frac{2}{3}\frac{2000}{13} = 100\ horas = 4\ días$$

The same equation can be used to reverse the process and determine the distance at which the storm originates. Two separate observations of the period of the cam waves are required for this. For example, suppose a smooth, regular cam sea with a period of 18 seconds moves in the first observation; 24 hours later, the period of the cam sea is 16 seconds. We can assume that the 18-second and 16-second cam sea left the storm at the same time, consider the symbol "t" to represent the time taken to move through the 18-second cam sea and solve the above equation for the distance, using the known data from both observations:

$$Distancia = \frac{3}{2} \times periodo \times tiempo\ de\ recorrido$$

$$= \frac{3}{2} \times 18\ t = \frac{3}{2} \times 16(t + 24\ horas)$$

Solving the last equality for "t":

$$18\,t = 16\,t + 16 \times 24$$

$$t = 8 \times 24\ horas = 8\ días$$

Therefore, the time taken for the 18-second cam sea was 8 days, and for the 16-second cam sea it was 9 days. Giving an appropriate value for "t" in the equation to find the distance, the distance to the storm would be 5184 nautical miles.

7.3. Effect on living organisms

The diversity of the components of marine flora and fauna is enormous and most of these organisms live in the surface layer or depend on it directly or indirectly.

The micro-organisms most influenced by their size are phyto- and zooplankton, which are driven by wind turbulence due to their low or no swimming capacity. Their abundance depends on the presence or absence of nutrients such as phosphates, nitrates, nitrites and silicates, which are products of the decomposition of organic matter found on the seabed. Due to the combined action of wind and Coriolis force, especially on the eastern sides of ocean basins, vertical upward movements (known as upwelling or upwelling) are generated, displacing masses of water from the sea floor to the ocean floor.

The sea floor to the surface, bringing with them these vital substances for phytoplankton, the first link in the marine trophic chain.

The distribution of marine plankton in turn conditions the distribution of other organisms at higher trophic levels. Each organism moves to where its food is found. Therefore, the areas with the greatest abundance of marine life are those areas that are characterised by the conditions for this abundance of nutrients to exist.

Another factor conditioning the distribution of living organisms is the climate. As we have already seen in previous chapters, there is a strong interaction between the atmosphere and the ocean. The climatic conditions influence the distribution of the oceans and vice versa and therefore condition the presence or absence of certain organisms. For fisheries this is very important because it increases the probability of success in catching fish resources.

By way of example, Peru, despite being located in a tropical zone, but because of the Peruvian current, which is part of the southern Pacific cell that is driven by the system of atmospheric currents, it carries waters that come from the subtropics very close to the subantarctic, which are characterised by their cold waters. This regime makes the Peruvian coast arid and cold. The predominant trade winds in conjunction with the Coriolis force produce the coastal upwelling that transfers nutrients from the shallow bottom of the continental shelf. Large schools of relatively cold-water pelagic fish such as anchovy and sardine predominate in this area.

It is also important for mariculture, because it allows to determine whether a certain area meets the requirements for the cultivation of any fishery resource, even if it is demersal, i.e. living on the bottom, as it still depends on the interaction with the atmosphere because these bottoms cannot be very deep for technical reasons and because of the high costs involved in cultivating at depths of 40 metres or more.

CHAPTER VIII

THE GENERAL SURFACE CIRCULATION OF THE OCEANS

The océano is a huge mass of water in constant motion following circulation patterns determined by a number of factors, but its main driving force is the winds. This is true especially for the surface layer. As already mentioned in previous chapters, the circulation patterns of both wind and ocean are very similar.

Both masses, air and water, shape the climatic conditions of our planet and therefore of human activity. Knowing the causes that determine the movement patterns of these masses is of crucial importance for the existence of human civilisation. For, the line of development is strongly dependent on the environment in which it develops. Natural forces are uncontrollable by man, and we have neither the science nor the technology to aspire to have any control over them. Our interest is mainly focused on knowing how it works, what variables are involved, etc., and for this purpose a great deal of money and resources are being invested, especially by developed countries. Phenomena of great magnitude like El Nino are barely understood so, as a species, we are on the defensive, seeking to predict their behaviour in order to mitigate their negative effects.

Having a clear understanding of the characteristics of the movements of the surface water masses of the oceans is important for the fisheries engineer because in this way we can predict, for a given area, where we will find the most favourable conditions or habitats for a particular species or group of hydrobiological species that we are interested in fishing or catching for the benefit of man.

In this chapter we will first briefly describe the global circulation of the Océanos, with its major circulation gyres. Then we will focus in more detail on the circulation of the Pacific Ocean in both hemispheres, with special emphasis on the Peruvian Sea. Of course, in the different currents mentioned we will always be referring to those found in the surface layer of the Océano. The importance of subsurface water circulation in the global climate is recognised, although it affects it indirectly, but it will not be dealt with here.

8.1. General circulation of the oceans

The surface ocean currents are the main vectors of movement capable of transporting large masses of water from one region to another, such movements are produced by various causes, of which mainly we have the action of the wind, influencing it also the rotation of the earth and the interference of the continents, its magnitude depends on the strength of the wind, and its speed is small compared to those of the wind.

In terms of their direction, they are referred to as the direction in which they are blowing, in contrast to winds, which are referred to as the direction in which they are blowing. The surface ocean circulation is the result of several processes, especially the force of the wind acting on the water surface and density differences. The circulation is very similar to the main wind belts on Earth. The main causes of ocean currents would be:

- Permanent winds blowing over the water surface (which produce friction and dragging of the surface oceanic water molecules).
- The influence of the disposition of continents and coastlines.

As we can see from the attached figure, in all the oceans, the general circulation is in closed cells. In the southern hemisphere there will be a cell in the Pacific, Atlantic and Indian Oceans which rotate counterclockwise. On the other hand, in the northern hemisphere in the Pacific Ocean there are two closed cells, the main one clockwise and the other, plus the northern one counterclockwise; in the Atlantic Ocean there is a closed cell and an open cell and in the Indian Ocean the current is irregular and does not form a closed cell due to the seasonal variation of the monsoon.

FIGURE 8.1. Source: **www.marviva.org/articulos/regatasmundo2.htm**

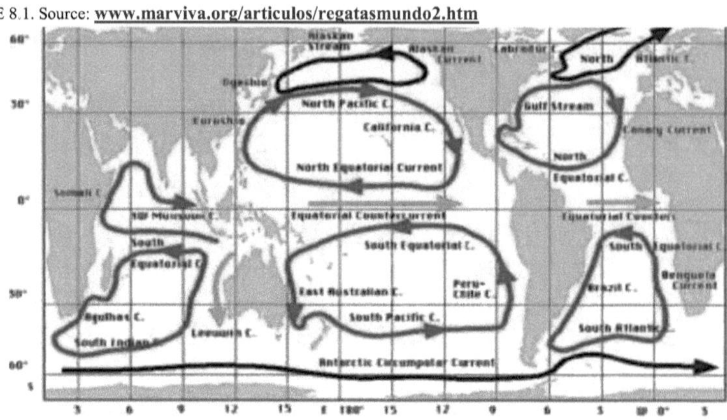

Despite the different shapes of the Atlantic, Indian and Pacific Oceans, they have similar surface current structures, dominated by a circulation (or gyre) of oceanic amplitude, with the currents being much stronger in the narrow regions close to the western borders.

The Gulf Stream in the North Atlantic and the Kuro-Shivo Current in the Pacific are the best known; the corresponding current in the Indian Ocean, the Somali Current, is complicated by the seasonal variation of the monsoon. Near the equator in all the oceans there are two westward-directed currents; in the Pacific, Indian and part of the Atlantic oceans, they are separated by an eastward-directed equatorial countercurrent.

In the Antarctic Ocean there is no continuous continental barrier (although the narrow Drake Passage may cause a similar effect) and the main surface current flows in a circle around the Earth in the Antarctic Circumpolar Current, in an easterly direction.

Published maps of surface ocean currents are based on average situations: in a particular case, the current may be very different, especially in currents such as the Gulf Stream with complicated swirling meanders and annular slopes. Large surface currents vary with wind and weather, but can be considered semi-permanent.

Oceanographer Munk proposes a general explanation for the formation of currents, based on the consideration of a theoretical ocean. If we consider 3 typical cases of increasing complexity, we can see the following.

Situation 1: This is a rectangular océano, whose main axis is N-S, with continental boundaries to the W and E. The océano is rotating and uniform zonal winds from the W are blowing over ël with a constant direction and intensity.

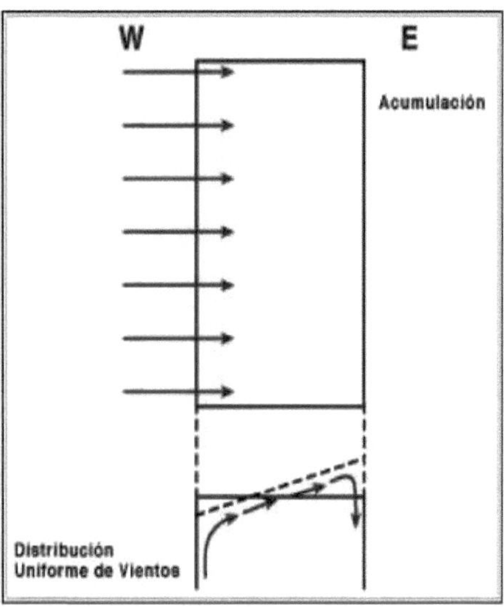

In this system there is a tendency for water to accumulate on the east side of the ocean, with a uniformly balanced slope due to the constant thrust of the wind.

In the vertical plane, a circulation is established with convergence on the east side and water sinking to a depth that is a function of the wind force and the vertical distribution of densities.

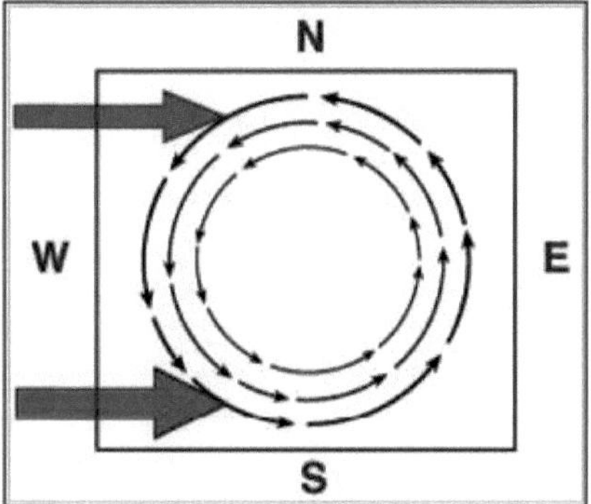

Situation 2: If in this same ocean, the winds have a constant direction, as in the previous case, but there is a variation in intensity with latitude, there will be asymmetry of the drag forces due to the wind.

In this case the slope will be steeper to the south than to the north, thus generating a horizontal movement in the east, carrying water from the high wind zone to the low wind zone.

in the east a horizontal movement carrying water from the area of strong winds to the area of weak winds. Due to the existence of continental Limits and the continuity of the winds, the movement will be rotational.

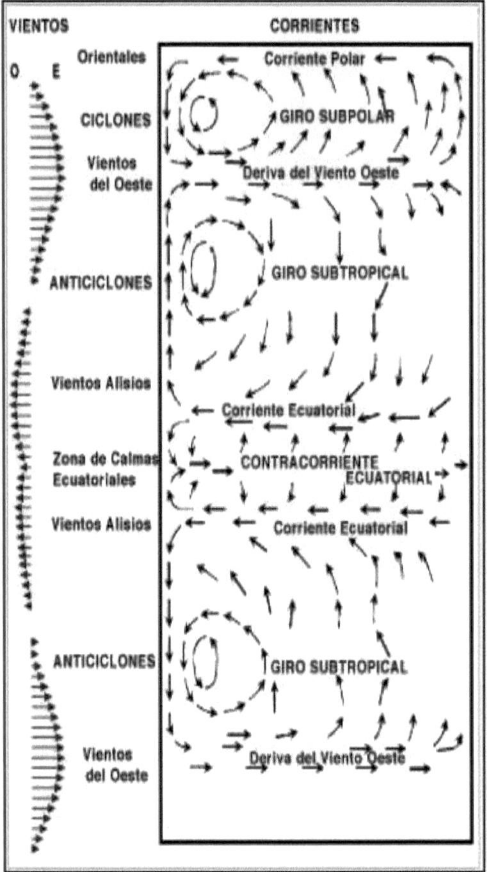

Situation 3: If the non-latitudinal winds of intensity, but also of direction, will be generated under the influence of these different forces, and in the presence of the east and west Limits of the oceans, an active circulation of cells only have variations.

horizontal with rotary movements (turns), in this case very close to reality.

A diagram of the circulation in an idealised océano subject only to the action of the winds is presented.

In conclusion: In this way the trade winds and the W winds determine in each hemisphere the circulation of the surface water of the ocean.

Each of the circuits has an equatorial branch in direction

E-W, and a subpolar branch running from W to E; the two are linked by circulations with a meridional component more or less parallel to the continents.

Between the two equatorial currents (separated by a doldrums zone), a reverse-direction equatorial countercurrent is formed (with a slope of 4 cm per 1000 km). It is a return current that carries through this calm zone a part of the water accumulated on the west side by the trade winds. A more important part of this water is taken out by the west meridian branch of the gyre circulation.

In the Northern Hemisphere, between the W winds and the polar circulation established by N and NE winds related to the polar anticyclone, a sub-polar cyclonic circulation appears.

Due to the rotation of the Earth to the E and also due to the variation of the intensity of the Coriolis effect with latitude, the centre of the gyres is shifted to the west side and the currents are stronger on the west side of the oceans than on the east side. Thus, in each gyre there is a persistent strong current on the west side and a compensating current in the central and eastern sector.

This can be recognised in the oceans; for example the strong Gulf Stream in the North Atlantic; the Kuro Shivo Current in the North Pacific (10 km/hour) while the California Current moves at a speed of less than 2 km/hour. In the South Atlantic the Brazil Current; the Agulhas Current in the Indian Ocean and the East Australian Current in the South Pacific.

8.2. The Pacific Ocean and its general circulation

The Pacific Ocean is the largest of the world's oceans in terms of both extent and depth. It covers more than a third of the Earth's surface and contains more than half of its volume of water. It is often artificially divided along the equator: the North Pacific and the South Pacific. It was discovered in 1513 by the Spaniard Vasco Nunez de Balboa, who called it the South Sea. The current name was given by the Portuguese navigator Ferdinand Magellan in 1520, during the first round-the-world voyage in the service of the Spanish Crown.

Boundaries and size. The Pacific Ocean is bounded on the east by the landmass of North, Central and South America; on the north by the Bering Strait; on the west by Asia and Australia; and on the south by Antarctica. To the southeast it is arbitrarily divided from the Atlantic Ocean by the Drake Passage at 68° W. To the southwest, the dividing line separating it from the Indian Ocean has not yet been officially established. In addition to the boundary seas that extend along its irregular western edge, the Pacific Ocean has an area of some 165 million km2, i.e. more than the entire continental surface. It has a maximum length of 15,500 km from the Bering Strait to the Antarctic, and a maximum length of 15,500 km from the Bering Strait to the Antarctic.

maximum width of about 17,700 km from Panama to the Malay Peninsula. Its average depth is 4,282 m, although the maximum known depth is 11,034 m in the Mariana Trench off Guam.

The main seas of the Pacific. The major seas of the Pacific present great differences in their extent, depths and topography causing variation in the deep circulation.

Data on the main seas of the Ocëano Patifico

Name of the seas	Area in thousands of km^2	Maximum depth in m	Some characteristics
Bering Sea	2292	4773	The shallow northern part
Sea of Okhotsk	1580	3846	Average depth > 1000 m
Sea of Japan	978	4246	Average depth 1752 m
Yellow Sea	417	106	Average depth: 40 m
East China Sea	752	2717	Shallow sea, except in the Riukiu Islands area.

South China Sea		5420	Predominant depths up to 200 m
Sea of Zulu	348	5119	
Celebes Sea	435	6220	
Band Sea	1227	7360	
Java Sea	480	89	Shallow sea.
Coral Sea	4791	9148	
Tasman Sea		5943	
Gulf of Alaska		3000	
Gulf of California	117	3127	

General currents of the Pacific Ocean. The general oceanic circulation of the waters of the Pacific surface map develops on a large scale, transports enormous volumes of water and covers large areas of the geographic space represented by the hydrosphere.

This circulation is based on (1) the action of the winds and (2) the perturbations that occur in the distribution of densities. Winds supply most of the external energy for the gënesis and maintenance of the movement of water transfer. The density gradient is, in turn, the main contributor of internal force to the circulation. But it should be remembered that an ocean current is not the product of a single external factor, but rather the result of the combined action of many factors, all of which vary in time and space.

The North Pacific current model consists of a motion, or circular system of two vortices (see Figure 8.1). In the northern hemisphere, the sub-arctic gyre is counter-clockwise, encompassing the westward subsidiary flow of the Alaskan Current and the eastward flow of the Sub-Arctic Current. However, the North Pacific water mass is dominated by a north central cell that flows clockwise and comprises the eastward-flowing North Pacific Current, the southeastward-flowing California Current, and the Kuro-Shivo Current, or Japan Current, which moves northward until it reaches the coast of Japan. The California Current is cold, broad and slow-moving, while the Kuro-Shivo Current is warm, narrow, fast, and similar to the Gulf Stream. Near the equator, at $5°$ N latitude, the eastward flow of the equatorial countercurrent separates the North and South Pacific current systems, although it transports most of its waters to the north equatorial current. The southern Pafic is dominated by the counter-clockwise movement of the south-central cëlula, which comprises the south equatorial current moving eastward and southward, the south Pafic current moving westward, and the Humboldt current moving northward in a direction parallel to South America. At the southern end is located the Antarctic Circumpolar Current, or westerly wind drift, the most important source of deep ocean circulation that circles the Earth and brings together the waters of the Pacific, Atlantic and Indian Oceans. This is the source of the large, cold Peru Current, which turns northward along the coast of South America and distributes its waters to the South Equatorial Current.

The Tropical Pacific Ocean zone. The Tropical Pacific zone contains the system of trans-oceanic equatorial currents and also part of the peripheral currents flowing towards the equator on the eastern side of this océano, between Cape Currents (Mexico) and Ecuador a more complicated circulation develops as the anticyclonic movements of the waters of both hemispheres do not extend to this part of the Eastern Pacific.

The surface water characteristics of the Tropical Pacific are typically tropical, except off the west coast of South America, where there are temperate waters, due to the coastal upwelling that brings cooler water to the surface from intermediate depths and the contribution from other latitudes made by the Peruvian Coastal Current.

Wind and current zones. The equator divides the Pacific Ocean into two parts of unequal shape and volume:

- The Northern Patifico, located north of the equator; and
- The Southern Patifico, south of the equator to the Antarctic Convergence.

In each part there are particular conditions that have an effect on the ocean circulation.

- The area of polar easterly winds.
- The westerly wind zone, -temperate zone.
- The trade zone.
- The area of equatorial westerlies, or calms.

The main surface currents of the Pacific Ocean flow according to the direction in which the winds blow. Thus:

1. In the area of the polar easterly winds; there are currents flowing from east to west.
2. In the area of the westerly winds of each hemisphere: the great eastward drift of the waters is developing. This movement reaches speeds of one nautical mile per hour and manages to cross the entire Pacific in a latitudinal direction.
3. In the trade wind zone of each hemisphere: there are the marginal currents flowing towards the equator and the equatorial currents flowing westwards,
4. In the area of the calms; there is the North Equatorial Countercurrent

However, it is worth mentioning that currents do not follow straight paths; rather, they flow in closed drums, sometimes -cum soleml, sometimes -contra soleml. By flowing within a closed circle, they form systems. Thus, the main surface currents maintained by the prevailing winds constitute two different systems:

- The system of equatorial currents and
- The system of polar currents in each hemisphere.

The two systems are linked by ocean currents that belong to both systems.

Current systems in the Pacific Ocean. Surface currents flow in the upper part of the ocean, between 0 and 100 to 200 m approximately. Caused by changes in atmospheric pressure, winds, waves and tides, we can distinguish:

- Surface currents caused by changes in atmospheric pressure.
- Surface currents produced and maintained by the wind,
- Surface currents caused by waves.
- Tidal streams that have periodical flow and
- Inertia currents.

When produced by the wind, currents reduce their speed with depth, and the depth at which the unmoving layer is located also influences their direction and speed. Finally, it is important to bear in mind that surface currents are influenced by water from other oceans and seas at contact zones and boundaries. The large contact zone between the Pacific Ocean and the Indian Ocean, on the one hand, and between the Pacific Ocean and the Antarctic Ocean, on the other hand, is of particular importance.

The Equatorial Current System. - The best known ones caused by plants are: The North Equatorial Current and the Surecutorial Current which flow westward, plus there is the North Equatorial Countercurrent which flows eastward.

The westward flowing surface currents are separated from each other by the North Equatorial Countercurrent which runs from west to east, having a more dëbilitating flow than those, their maximum velocity and their volume and extent vary with changes in wind regimes.

The North Pacific Gyre and its components comprise the following currents: the North Equatorial Current, the Kuroshio Current, the North Pacific Current and the California Current.

The cyclonic gyre of the northern hemisphere consists of the Alaskan Current, which then turns westward and passes near the Aleutian Islands, where it is called the Aleutian Current, flows from east to west and partly reaches the Kamchatka Peninsula. This water joins the waters moving southwards, giving rise to the Oyashio Current, which moves its waters from north to south until it meets the Kuroshio Current off Japan. Part of its cold water sinks at about 35°N latitude and continues to flow southwards but as a bottom current. Another part turns eastward and crosses the ocean together with the warmer waters of the Kuroshio and is called the North Pacific Current. This current closes the cyclonic gyre of the waters in the northern part of the ocean.

The *North Pacific Current* is driven by westerly winds. It is much less developed than the South Pacific Current, although it plays an important role in the two gyres in the North Pacific.

The *Alaska Current* is the northern branch of the North Pacific Current.

The *Aleutian Current* flows westward as a continuation of the Alaskan Current; it has several branches between the islands of the Alaskan Archipelago that mark its course. Part of its waters crosses the Bering Strait and enters the Arctic.

The *Oyashio Current* flows along the coast of Eastern Siberia and the Curile Islands to the southwest, carrying cold water with ice. At the meeting place with the warm Kurosthian Current, its waters turn southward and then turn into the cold water of the Kurosthian Current, to the east. Some of this water merges with the Kuroshio Current and some with the Aleutian Current. From the point where the eastward shift in orientation occurs and the waters of the Kuroshio Current meet, the waters moving horizontally eastward form the great North Pacific Drift. This closes the water circuit of this large region of the North Pacific.

In the South Pacific, there is an anticyclonic and a cyclonic gyre of the surface waters. The South Pacific anticyclonic gyre is located between the calm zone and the subtropical zone. It is composed of the East Australian Current, the South Pacific Current and the Peru Current, which flows into the South Equatorial Current.

The cyclonic gyre is concentrated in the Antarctic polar zone and part of the subantarctic zone and therefore belongs to the Antarctic Ocean. Its components are mentioned here to complete the general picture of the major circulations in this part of the geographic space. They are the Antarctic Circumpolar Current which moves eastward around the Antarctic continent and the Antarctic Current which flows from east to west near the coast of the white continent.

The temperature and salinity of the surface waters in the currents of the Padfic. The temperatures and salinities of the surface layer of the Océano Patifico by streams, for the months of February, March (II-III) and August, September (VIII-IX), according to Bruns 1958.

Name of the currents	Months	Temperature	Salinity ^
Alaska Current	II - III	+ 2 a +3	32-33
	VIII - IX	+13 a+10	
Oyashio Current	II - III	+1 a 0	31-33
	VIII - IX	+7 a +10	

North Patifico Current	II - III	+10 a+20	35-33
	VIII - IX	+26 a+13	
California Current.	II - III	+10 a+15	32.5-34
	VIII - IX	+10 a+20	
North Equatorial Current	II - III	+18 a+27	34-35.5
	VIII - IX	+25 a +28	
Kuroshio Current	II - III	+17 a+10	34.5 - 33
	VIII - IX	+28 a +20	
North Equatorial Countercurrent	II - III	+28 a +25	34 - 34.5
	VIII - IX	+28 a +26	
Southern Patifico			
Surecuatorial Current	II - III	+24 a +28	35-36
	VIII - IX	+20 a +28	
Eastern Australia Current	II - III	+26 a+15	35-35.5
	VIII - IX	+20 a+10	
South Pacific Current	II - III	+5 a +15	33.6-35
	VIII - IX	+1 a +10	
Peru Current	II - III	+20 a +25	34-9-35.5
	VIII - IX	+17 a+20	
Antarctic Circumpolar Current	II - III	-2 a +2	35.5
	VIII - IX	-1	

Classification of the Patifico currents

Transoceanic circulation subdivided into equatorial circulation, temperate circulation and polar circulation; and marginal or periphytic circulation. The equatorial circulation consists of zonal currents, of which two flow westward, namely the Surecuatorial and the North Equatorial currents, and a North Equatorial countercurrent moves eastward.

The first two are broad currents resulting from the action of the trade winds, while the North Equatorial Countercurrent is conceived as a compensating flow originating from a larger accumulation of water on the western side of the Octiane.

The marginal circulation develops in the region adjacent to the coasts of the continents, having as representative elements the coastal or peripheral currents that generally have their flow parallel to the coast or slightly deviated.

Marginal currents exist both on the western and eastern side of the Pacific (also in the Atlantic and Indian Ocean), flowing longitudinally, generally along the west coast and towards the equator off the eastern coast of the continents. But there are also marginal currents and countercurrents that flow in a closed circle and in the opposite direction, which introduces some complications into the overall picture of the circulation.

The causes of the marginal currents are: winds, waves and undertow. Some peripheral currents are very important, as is the case, for example, with the Peruvian Coastal Current which maintains its flow under the impetus of the SE trade wind, and the California Current off the west coast of North America which has the NE trade wind as the source of its driving force.

On the western side of the Pacific there is a greater concentration of the flow of currents. AШ is the northward flowing Kuroshio Current and the southward flowing East Australian

Current. On the eastern side of the Pacific, the marginal currents do not have a concentration of flow equal to that on the western side. These currents are generally wide and carry large volumes of water, but are relatively slow.

All marginal currents on the eastern side of the Pacific (and the Atlantic) flow along the eastern side of the great subtropical gyre of waters in the respective region. In the Southern Hemisphere, the largest spatially extensive marginal surface currents on the eastern side of the Ocetiane flow equatorward, moving their waters from higher latitudes to lower latitudes, and can reach widths of almost 1000 km.

All the eastern marginal currents are relatively slow, having an average velocity of less than one knot. All these currents are shallow. Their depth is less than 500 m. Their transport is of the order of 15 x 106 m^3 /s. All these characteristics have an effect on the dynamics of the waters in the equatorial region and have long been one of the exciting topics of dynamic oceanography.

Equatorial current system

The North Equatorial Current. The NE trade winds cause the waters of the North Equatorial Current to flow westwards, which represents the essential part of the circulation of warm waters in the North Pacific. It is a zonal current, belonging to the equatorial current system. Its waters cross the entire Pacific from east to west with increasing velocity in the same direction. North of Mindanao Island, part of its waters turn southward and flow in that direction throughout the year, feeding the North Equatorial Countercurrent. Another part flows northwards and contributes to the formation of the Kuroshio Current.

Between the latitudes of 10°N and 20°N, it oscillates in space with the seasons of the year. After crossing the Ocëano, it bifurcates and its main branches extend to the south and north.

Dimensions: The average width of the current is about 650 nautical miles. The total width of the current varies along the different lengths.

Structure: The surface layer, between the surface and the thermocline, has a thickness of 50150 m. temperature of 24-30°C and salinity of less than 34.4%; the discontinuity layer, which extends up to the 10°C isotherm approximately and is closer to the surface on the southern margin of the current than on the northern margin; the lower layer, with a temperature of less than 10°C, sometimes called the "free layer", has a salinity of less than 34.6%.

Speed: Almost constant 0.3 knots throughout the year.

Direction: It flows from east to west, but changes its general direction at the western end, owing to the presence of numerous islands in its path and to the action of monsoons blowing from almost opposite directions in successive seasons.

Transport - The North Equatorial current moves 30 to 45 million m^3 /s, volume according to some authors.

Characteristics: Modified from east to west by mixing with other waters, the surface layer of the stream is of relatively low salinity. This property has a maximum in the upper part of the thermal discontinuity layer. The high salinity has its origins in the subtropical water of the North Pacific.

The low salinity of the upper layer is generally found on the southern side of the Current. The lowest salinity (> 34.0%) found in the lower layer is attributed to intermediate water from the North Pacific. Salinities somewhat higher than 35.2% are found in the western part of this stream and are attributed to the input of subtropical water from the North Pacific. The water temperature of this stream is 26° to 27°C in the eastern part and 28-29°C in the western part.

Role and importance: It is the main link between the eastern and western sides of the Equatorial Pacific. It contributes largely to the mixing between the waters it receives from the California Current and the other waters of the central region of the North Pacific. This mixing also includes the waters of the equatorial countercurrents.

The Sub-equatorial Current. Winds and zonal surface currents are not symmetrically distributed on either side of the equator, as it is the SE trade winds that prevail at the equator. Because of this, equatorial surface waters flow westward on both sides of the equator as the - southern equatorial current-.

Origin: It is formed in the Eastern Pacific region.

Position: It is found north of the South Pacific Subtropical anticyclone, where it flows permanently westward under the impulse of the SE trade wind. It is present on both sides of the equator.

Boundaries: The northern boundary is formed by the North Equatorial Countercurrent and is normally between latitudes 3°N and 4°N. Its southern boundary is very variable in the different longitudes, modifying its position with the seasons.

Dimensions: It is more developed until September, its strength decreases in October, from November to February it widens and flows between 1°N and 4°N.

Stability: Not established, as it depends on the variable intensity of the SE trade winds in the eastern Pacific and the monsoon regime in the western Pacific.

Velocity - Flows most strongly from June to August, when it reaches a velocity of 100 cm/s.

Direction: The westward flow loses its intensity and frequently changes its direction.

They have relatively low temperatures on the eastern side due to the effect of the waters of the Peruvian Current and local upwelling in the Galapagos area. The temperature and salinity of the waters increase towards the west.

Role and importance: The Equatorial Current displaces huge volumes of water in the equatorial region and thus contributes to the westward extension of the favourable upwelling effects of the Peruvian and Galapagos upwelling waters.

Current systems in the North Pacific

They showed that in the North Pacific there are several systems of currents, connected to each other.

- The Kurostium Current system.
- The drift of the waters of the Northern Patifico
- The California Current System
- Ocean currents in the Subarctic Region of the Pacific and the Pacific Ocean.
- Currents in the Bering State.

The Kuroshio current system. On the western side of the North Pacific there is a complex system of currents consisting of several channels, eddies and counter-currents; it spans different seas off Asia and has as its trunk the Kuroshio, one of the largest warm ocean currents in the world, known for centuries to fishermen and sailors in the region.

Background. This stream first appeared on a geographical map in the 17th century. Berghaus (1837) called it -Japan Currentll instead of Kurosdo, but this name change did not succeed. Scientific investigation of this ocean current began in 1893, when flasks were dropped into the water to observe the direction and speed of the flow of the water. Today, the Kurosdo and its hydroclimate are considered to play the same role in the North Pacific as the Gulf Stream does in the North Atlantic.

Description. Kurioshio is a well-defined marginal current with strong, concentrated flow that has the shape of a band and moves warm tropical waters from Formosa to the northern edge of the subtropical region, passing some distance off the southeastern coast of Japan. It is usually accompanied by strong horizontal temperature gradients indicating the presence of oceanic fronts.

Origin. Because it rises in the region situated to the southeast of the island of Formosa and east of Luzon, receiving waters from this current.

Position. The Kuroshio Current flows northwards, practically bordering the Japanese islands.

Width. The Kurioshio Current has variable dimensions. In the summer it can reach a width of about 300 nautical miles. In the winter, when a branch of the Kurioshio passes through the Luzon Strait and enters the South China Sea, where it subdivides in turn.

Speed. The velocity is 1 to 3 knots. The flow of the warm Kuroshio Current varies in space and time, in fact, between 1 and 5 knots, in relation to changes in the atmospheric pressure field. The maximum velocity on its axis is proportional to the vertical thermal gradient of the water column between 0 - 200m. In the summer it reaches speeds of 24-36 nautical miles/24 hours, increasing to 36-48 nautical miles/24 hours off the eastern side of the Riu-Kiu Islands; at 30°N, it reaches 48-56 nautical miles/24 hours. Off the Boso Peninsula, at 36°N latitude, when turning east, the current speed starts to decrease from about 50 nautical miles/day to about 24 miles/day. *Transport*. Generally, the volume of water transported by the Kuroshio increases from south to north, since as it moves northward it draws water from the large eddy to its right and the coastal seas to its left. Authors who have studied this aspect of the Kuroshio admit that the volume transported varies between 30 and 50 x 106 m /s.[3]

Characteristics. The waters of the Kuroshio mainstream have limited transparency to 25.40 m. The salinity is lower than the Pacific average 34.9°/00, however, it is high for the area it crosses, being higher than 34.5°/00 in the core of the stream, the salinity is normally between 34.8°/00 and 35.1°/00. The dissolved oxygen content is about 5 to 5.5 cm^3 /L. All these properties of the Kuroshio Current change with distance travelled as coastal waters from different seas are added to it and mixed with it.

Role and importance. As they move northward, the waters of the Kuroshio lose some of their warmth; nevertheless they retain a sufficiently high temperature to have a benign effect on the general conditions of the region. Thus the southern islands of Japan which are under the effect of the warm waters of this current acquire an almost tropical climate. Meanwhile, the northern islands of the respective country which are under the influence of the cold Oyashio and Sakhlin currents have an arctic character.

The drift of the waters or North Pacific Currents. The waters of the Oyashio Current and the Kuroshio Current flow side by side slowly from west to east across the Pacific. In this two-year-long movement to the region off North America, the waters of both currents lose their initial characteristics and, by the addition of other waters, gradually widen. From the 138°W meridian eastwards, they are called the North Pacific Current.

In reality, it is the movement of the surface waters under the impulse of the westerly winds. Its speed is 0.50 min/h. This current has its origins to the east of the confluence area of the two currents mentioned above and its terminus off the coast of North America, where it is subdivided into two branches: one running northwards, off the state of Washington, and forming the Alaskan Current; the other running southwards as the California Current. This branch establishes the junction with the North Equatorial Current, to which we draw our attention in the following lines.

The California Current System. In front of California there is a -transition region between the northern bellows zone and the equatorial zone in the south. Here the California Current system develops its flow, which includes the -Californial Current, the -Submarine Countercurrent and the -Davidson Countercurrent.

Background. One of the regions where ocean currents have been studied with the greatest dedication is off California. The series of scientific studies of these currents was initiated by Tthorade (1909), and was pursued with greater enthusiasm between 1931-1941. Sverdrup and Fleming (1941), Tibby (1941) and others contributed valuable insights into the problem of circulation in the region. These efforts were later joined by Wooster and Cromwell (1958), Reid (1961), Roden (1962), Wyrtki (1965) and others. The California Current is the largest of the marginal currents of the Northeast Pacific, flowing southward to replace the westward-moving waters under the impetus of the NE trade winds. The current receives also the cooler waters off the coast of California, especially in the summer course.

Origin. Shortly before reaching the American coast, the North Pacific Current splits into two branches: one goes northwards to form the Alaska Current and the other goes southwards to form the California Current. This is a perifēricall North Pacific Current that transports relatively cold waters southward. This current flows off the west coast of North America in a generally southeasterly direction from about 48°N to 23°N latitude. It is approximately 200 m (between 0-200 m). Seasonal wind variations and upwelling processes in the coastal area introduce important changes in the flow velocities of the California Current during the different months of the year. In the area where the current turns westward into the North Equatorial Current, the average velocities remain very low, but almost uniform at 0.3 knots throughout the year. Despite being very wide, the California Current carries only about 10 million m^3 /s between the surface and the 1500 decibar level.

Characteristics: The California Current consists of subarctic water, central water and equatorial water intermingled.

Salinity. Relatively high salinity waters enter the region in the summer (May to September) and winter (December and January).

Temperature. The current has relatively low temperatures; this is contributed to by coastal upwelling, which starts in March and continues until July. Spring temperatures are lower than winter temperatures in all areas of intense upwelling. From these upwellings, tongues of cold water extend southwards, separated from each other by tongues of warmer water that move towards the coast in a generally northerly direction.

The California Current looks like a mirror image of the Peru Current. Both are marginal currents, both belong to an anticyclonic gyre of the waters and both are directed towards the equator, being connected to the coastal upwelling areas. What distinguishes them is the latitude at which they originate and the latitude at which they incorporate their waters into one of the equatorial currents.

The Davidson Countercurrent. American authors mention in their studies the -Davidson Currentll. But in reality it is a counter-current subsurface, because it flows in the opposite direction to the NE trade wind and the California Current. As coastal upwelling ceases in the California region during the autumn, a countercurrent develops on the eastern flank of the California Current; it flows northward during the winter.

Course. The Davidson Countercurrent flows from south to north within the coastal area of California, reaching its greatest development in the months of November to January, when it appears at the surface.

Width. It is variable and can reach up to 165 nautical miles.

The California Countercurrent. During the upwelling temperature a subsurface countercurrent develops off California. According to some authors (Sverdrup, Jonson and Fleming), this countercurrent would be a subsurface manifestation of the deeper countercurrent that can occur only in certain seasons, when the wind weakens.

Course. The California Countercurrent flows below the 200 m depth near the west coast of the United States of North America, moving equatorial waters northward under the California Current.

Width. It is 40 to 50 nautical miles at latitude 36°N.

Speed. 0.5 knots off Washington State and 0.2 knots off California. *Direction*. The countercurrent reaches its full development in January (northern hemisphere winter); then, at a depth of 250 m., its general flow is northward, up to 100 miles offshore, then northwestward. This flow presents almost periodical fluctuations with diurnal and semi-diurnal periods.

Circulation in the Gulf of California and its vicinity.

(1) **The circulation in the Gulf of California** is an exception to the circulation in the rest of the ocean between the same latitudes. The water mass in this gulf comes from the Equatorial Pacific and maintains its characteristics below the thermocline, while in the surface layer it is modified by intense evaporation and mixing with traitii water by the California Current.

(2) **The circulation in the vicinity of the Gulf of California** is affected by water transported southward by the California Current and by water off Mexico and Central America. It is characterised by low temperature, low salinity and relatively high dissolved oxygen content. It has very high surface temperature and very low oxygen content below the thermocline. Between the two waters of different characteristics, a distinct transition region between 17°N and 20°N is required. This is the front between the California Current and the surface water of the Eastern Tropical Pacific, which is between 0-50m.

Ocean currents in the sub-arctic region of the Pacific. Pacific sub-arcticoll is defined as a region in which there is a distinct halocline separating a less salty upper layer from a saltier lower layer.

Subarctic features are limited to the upper and halocline layer. The upper layer extends to a depth of 85 (+75) to 200 (+50) m and shows an increase in salinity to 33.8 + 0.1%. The low salinity of the upper layer and halocline is an effect of the excess of precipitation over evaporation in this region, which plays an important role in the subarctic circulation. In the subarctic pallid climate, a cyclonic circulation develops under the impulse of the winds in which it participates:

The Alaska Current. It flows around the Gulf of Alaska, but part of its waters crosses the Pacific to the west along the southern Aleutian Island chain and disperses into the Bering Sea. This westward flowing part is called the Alaska Current. The other part of the waters flowing around the Gulf of the same name is the Alaskan Gyre.

Origin. The Alaska Current has its origin in the terminal area of the North Pacific Drift.

Velocity. On the western side of the subarctic Pacific it flows with velocities of 80 cm/sec for the surface and 60 cm/sec for the 300 m depth. Therefore, it should be noted that there are differences in velocity at different depths.

Direction. Parachute float measurements indicated that the direction of flow of the Alaskan Current is westward.

Characteristics. The Alaskan Current is characterised by a temperature above 4°C and salinity below 32.6% at the surface. The 4°C isotherm deepens to 400 m. off the Aleutian threshold, at full current, and rises steeply to 25 m. above this threshold where the current is absent. Salinity remains low over the entire area of the Current.

Role and importance. The waters of the Alaskan Current are generally warmer than the surrounding waters and therefore exert a softening influence on the entire region. Their presence is a source of attraction for many fish in the Subarctic region.

The Oyashio Current is peripheral and moves cold water southwards and then eastwards.

Origin. The current originates in the Bering Sea from the waters around the Bering Sea and flows in an anti-clockwise direction.

Velocity. Generally, the velocity of this current varies with the seasons, but there are also variations from year to year.

Direction. The Oyashio flows south and southeast.

Transport. Carries arctic waters of low temperature and salinity to Hokkaido. *Characteristics*. This marginal current remains under the influence of the continental climate of the atmosphere with which it is in contact. Its waters have low temperature and relatively low salinity. The temperature is 1°C to 2°C and the salinity is 35.5%. In the summer there is a warming of the surface waters to a depth of about 25 m.

Role and Importance. The cold arctic-like waters of the Oyashio Current impose arctic conditions on the island of Hokkaido (Japan) and the confluence region with the Kuroshio Current, located between 35°N and 40°N. The north-south oscillations of the confluence area have considerable effects on agriculture in northern Japan. The Japanese make extensive use of the knowledge of this fish behaviour, concentrating a large part of their fishing activities in the respective area.

The Sakhalin Current. It flows in the Sea of Okhotsk off the eastern coast of the island of the same name and carries cold water of very low salinity.

Origin. This current originates in the sub-Arctic zone as a component part of the swirling movement of surface waters between Kamchatka and Sakhalin Island.

Dimensions. The Sakhalin Current is typically shallow; its waters are present only between 0-100m.

Transport. The volume of water transported by this stream is unknown, but it is accepted that it changes at the beginning of each winter.

Characteristics. The Sakhalin Current is cold and is characterised by its almost uniform temperature between 0-100 m and its chlorinity (Cl = 17.80 in the summer) which is due to the enormous amounts of ice that form in the Sea of Okhotsk, where they reach a thickness of about 1 m. during each winter.

Important Role. It plays an important role in the drifting of ice along the east coast of the island of the same name. It moves large amounts of ice towards the north coast of Hokkaido Island, affecting the life and industries of the region.

The currents of the Bering Strait. The Bering Strait connects the waters of the Arctic with those of the Bering Sea. The Bering Strait is characterised by its small width and shallow depth. Both characteristics contribute to the increase in the speed of the surface currents.

Atmospheric circulation in this region is characterised by air currents moving from north to south for much of the year with maximum transport in February. The variation in the intensity of the winds is reflected in the circulation of the currents crossing the Bering Strait:

The Bering Current. This name is used here for the first time to replace the name -Beringl Sea Current- used by other authors.

Origin. This current originates in the Bering Sea and plays the role of contributing to the exchange of water between the North Pacific and the Arctic.

Velocity. The current flows throughout the anus, although with varying intensity. Its speed can reach up to 40 cm/s.

Direction. The Current moves from south to north, crossing the Bering Strait on the eastern side.

Transport. This current provides 22% of the volume of water supplying the Arctic. The minimum contribution occurs during the winter and early spring and the maximum in July and August, i.e. during the northern hemisphere summer. The Bering Strait offers favourable conditions for the study of variations in water and heat transport in different years.

Characteristics. Relatively high temperature and high silica content.

Role and importance. Due to the particular characteristics of its waters, the Bering Current exerts its greatest influence on the hydrological regime of the Arctic Basin in the SE Siberian Sea. In the years when the waters of the current manage to advance to the pole, the thickness of the ice sheet changes, and in general the conditions in the upper 50-100 m layer.

The Polar Current. At the western end of the Bering Strait flows an irregular current called the Polar Current. It is at its strongest in the autumn. Its origin is in the Arctic region. For its characteristics and role in the transport of ice.

Pacific ocean currents in the southern hemisphere. The horizontal currents of the South Pacific are dynamic elements that are directly related to the great wind systems existing in the southern hemisphere and to the general movement of the waters that contribute to their origin. The part of the Pacific Ocean which is situated in the Southern Hemisphere, under the dominance of 2 wind systems: The system of westerly winds, called -Bramadorl and the system of southeasterly tropical winds, called -Alisiosl. The action of the winds on the surface of the ocean and their effect on the first layers of water is of particular importance for the oceanic circulation.

Each wind system lends part of its force to the origin and maintenance of ocean currents. The stronger the winds, the greater their influence on the surface movement of the waters and the greater the influence of the general ocean circulation on the local situation in this surface movement. The changes which cause, in addition, variations in the intensity of the vertical movement of the waters along the margin of the continents are best illustrated by the equatorial upwelling.

The main oceanic current systems in the South Pacific are as follows: The East Australian Current System, the Great South Pacific Transoceanic Drift and the Peru Current System.

The East Australian Current system. A complex circulation is developing off eastern Australia. The oceanic currents and counter currents form the East Australian Current system, which comprises: A southward-moving branch between 28 S-33 S within 110 km from the edge of the continental shelf; A strong, offshore branch, 220-320 km, which also flows southward; A strong, offshore branch, 220-320 km, which also flows southward; A strong, offshore branch, 220-320 km, which also flows southward. Another strong offshore branch, 220-320 km, which also flows southward; an anticyclone south of 39 S, eddies and currents associated with them, subsurface currents. The following is a description of the main currents in the system.

The Western Australia Current. It is a marginal, wind-driven, southward flow near the edge of the Australian continental shelf between latitudes 27°S-33°S throughout the year with varying intensity. This current has two branches, an inner and an outer branch. The lower branch flows southward close to the coast, moving away from the coast between latitudes 33°S - 34°S, heading north or northeast. The outer branch heads north and flows as a - Contracorrientell.

Origin. The warm waters of this current feed in the western part of its course into an ocean current that flows southwards off Eastern Austria, which is why it is called the East Australian Current.

Position. The inland branch of the southward flowing East Australia Current lies within 110 km of the edge of the continental shelf.

Dimensions. The southward-flowing branch of the East Australia Current is 100 to 150 km wide, but the core varies between 40 and 80 km.

Stability. Australia's surface current is characterised by its lack of constancy. It flows southward in certain seasons and northward in other seasons.

Velocity. The intensity of this current is greatest from December to March (summer), with speeds of about 5 cm/s. At its core the current reaches speeds of 50 cm/s. The maximum speed is 2 knots at latitude 30°S.

Direction. The inner branch flows southwards, the outer branch moves in the opposite direction, i.e. northwards.

Characteristics. The surface temperature shows a change from 1.5° to 3.5°C. The vertical temperature distribution varies with depth.

Importance. The inland current of eastern Australia moves very warm water southwards so that many tropical animals can live here.

The ***East Australian Counter Current***. Or return current is part of the East Australian Current. The countercurrent occupies a position always east of the inland branch of the East Australian Current system.

The ***Tasman Current***. In the Tasman Sea, between latitudes 30°S and 35°S, there is a surface current that flows eastward across northern New Zealand. When it reaches the area of influence of the Bramadors, its waters join the South Pacific Current.

The ***East New Zealand*** Current. It flows to Tasmania Island in the form of an eddy system.

The great trans-oceanic drift of waters in the Southern Pacific. Between the South Pacific Anticyclone and the Antarctic continent lie the Subantarctic region and the Antarctic region, separated from each other by the Antarctic Convergence. Here we can distinguish 3 movements of the surface waters: the great trans-oceanic drift of the South Pacific, the Antarctic Circumpolar Current and the South Antarctic Current (westward).

The great trans-oceanic South Pacific Drift, also known as the South Pacific Current, is generated by the upwelling and develops between latitude 35°S in the north and the Antarctic Convergence in the south, whose position is between 47°S and 63°S, varying in each longitude. The region of the South Pacific Drift is divided into two unequal parts, with the Antarctic Convergence zone as the dividing zone, whose geographical position is from south to north and vice versa with the seasons of the year. The movement of the surface waters towards the east allows us to recognise two parts: a northerly one between 35°S and 40°S and a southerly one between 40°S and 63°S. The northerly part belongs to the Subtropical anti-clonic movement of the South Pacific waters, while the southerly part is a component of the Antarctic Circumpolar Current. The large South Pacific Current moves huge volumes of water from west to east, although it is dëbil and highly variable.

Origin. It originates in eastern and southern Australia. In eastern Australia it is fed by the East Australian and Tasmanian currents. In the south it receives large volumes of water from the

Indian Ocean.

Borders. On the western side of the Pacific there is no boundary, only Australia can be constituted, in part. In the north and south, the subtropical and Antarctic convergences can form a boundary. In the east, the boundary is represented by the South American Current which extends to latitude 56°S. Part of the water is oriented to the south and bends around Cape Horn under the name of the Cape Horn Current; the other part is oriented to the north, giving rise to the Peruvian Current.

Dimensions. The east can be identified from the surface to a depth of about 500 m in the area between 30°S and 40°S to the south, the situation is complicated by subsidence movements of some of the water which becomes denser as it mixes with Antarctic waters.

Stability. The surface waters from the South Pacific to the east show seasonal variations and also major irregularities in the same way as the intensity of the wind. From time to time the waters of the South Pacific turn northward earlier than usual, which has an impact on all other surface currents and their boundaries.

Velocity. Most of the water in the Southern Pacific moves eastwards with velocities of 2 to 9 cm/s at the surface increasing to 12 cm/s. Velocity decreases towards greater depths, being very small below 300 m.

Direction. The waters generally move eastwards, but at 35°S latitude they appear to veer westwards; only below a depth of 400m.

Characteristics. The waters of the South Pacific have characteristics of temperate zone waters, but at its southern boundary it receives polar water from the Antarctic which, driven by the winds, manages to cross the Antarctic Convergence. In the eastern part, the Cape Horn Current has temperatures of 5°C to 6°C in November and December throughout South America. Further south, towards the Antarctic convergence, the temperature drops rapidly to about 2°C in the summer.

The ***Antarctic Circumpolar Current***. It is the only surface ocean current that flows freely around a continent. The geographic space in which it flows does not present any natural obstacles to its progress, so that it manages to cross all longitudes and go around the Antarctic continent. Between the current and the South Pacific Current there are eddies and countercurrents, which play the most important role in the narrow convergence zone; this is located in the latitude of 61°S and 62°S in the longitudes of 80°W and 90°W, in which area the current regime is rather uncertain and the speed varies greatly. It reaches 0.5 to 1.0 knot on the southern Chilean coast, while south of the convergence it is more than 0.5 knot. Thus, it is worth mentioning that waters of the circumpolar current penetrate the South Pacific, as in the other oceans, but only in the former do they reach low latitudes in a relatively unmixed condition.

The ***South Antarctic Current***. In the south, near the Antarctic continent, easterly winds blow almost all year round. The surface currents that form there flow westward near the shore of the continent, carrying its cold waters. The South Antarctic Current is strong in the west, being separated from the Antarctic Circumpolar Current by the Antarctic Divergence II, located at latitudes 70°S and 72°S over longitudes 80° and 90°W. North of it, westerly winds blow strongly. ***Role and importance***. The South Seantarctic Current facilitates the penetration of ships to the south; its warm waters are the access route for many pelagic and phytoplanktonic organisms to the south of the Weddel Sea. The South Seantarctic Current frequently carries large ice floes that have broken off from the ice shelf.

Currents In The Gulf Of Guayaquil. The Gulf of Guayaquil extends about 180 km inland

and reaches a width of 100 km between Punta Salina and Punta Santa Elena. The Gulf region is under the trade wind regime which has an annual cycle. They blow with greater intensity during the winter season, weakening in the summer. According to the changes in the atmospheric circulation there are changes in the circulation of the waters, which are greater in the stages of the winter wind regime than in the summer wind regime and in this region.

The oceanographic *characteristics* of the Gulf of Guayaquil. The waters of the Guayas and Tumbes rivers mix with the marine waters, giving rise to waters of low salinity (lower than 34.0°/00) and high temperature (higher than 20°C). This water is light and occupies the upper part of the sea, moving towards the west. In the outer gulf hemisphere, they meet the colder waters coming from the north-west coast of Peru. The contact zone between the waters is usually clearly distinguishable on calm sea days both by the difference in the colour of the water and by the presence of an area of scarps that extends to the northwest, disappearing over the horizon.

Structure in the Gulf. Below the light surface water there is a pronounced temperature gradient which represents the most prominent features. The layer with this temperature discontinuity deepens towards the equatorial coast and is usually of obvious discontinuity also for the vertical distribution of the other water properties. Within this discontinuity layer the water flows slowly towards the coast.

The surface current moves normally to the SW in October and November, the sea temperature between Punta Santa Elena and the Galapagos Islands in these months from 22° to 23.5°C. In November a warm water tongue appears with a temperature of 23.5°C to 25.1°C and salinity of 33°/00. This indicates that the surface water comes from an active warm water movement. In December-February there is usually an invasion of warm water from the east of Galapagos which affects the oceanographic conditions of the region.

The Peru Current System. Winds blowing off Peru parallel to the coast and oriented towards the equator lead to a westward transport of water in a divergent movement with respect to the west coast of South America. The waters that take part in this movement belong to the Peru Current system.

This system is located on the eastern side of the subtropical anticyclone. Its surface branches flow northwestward in the area of the southeasterly trade wind and join the South Equatorial Current. The waters moving off the west coast of South America, which belong to the anticyclonic gyre, are called the Perul Current. It is actually a system of currents consisting of:

- The branches of the Peruvian surface current flowing northwestwards.
- Subsurface countercurrents that may also occur transiently at the sea surface.
- Shallow, irregular countercurrents.
- Tidal currents.
- The whirlpools.
- The outcrop.

The Peruvian Current has: the Peruvian Coastal Current and the Peruvian Oceanic Current which are formed in the area of the greatest local oscillations of velocities.

The *Peruvian Coastal Current*. This name is applied to the system of coastal currents which have often been associated with the name -Humboldt's Current.

The *Peruvian Oceanic Current*. This name will be retained for the offshore waters that share the northward movement of the anticyclonic circulation but are different in composition.

The two names will also be kept in our description, although the waters of both branches merge in the winter, when the strong trade winds force the Peru Countercurrent to remain below the sea surface, at a certain depth. The Peruvian Coastal Current. It originates in the region off the west coast of South America. In the respective area it is located in summer between 40°S and 41°S and in winter between 32°S and 33°S.

Position. The Peruvian Coastal Current overlaps slightly with the coastal upwelling region in northern Chile and along the coast of Peru. In the northwest of Peru, its waters turn towards the Galapagos Islands; beyond this archipelago the current delivers its waters to the Surecuatorian Current which flows on both sides of the geographic equator.

Borders. They are as follows:

• *Meteorological limits*. Perhaps the most important is the condensation of clouds over the cooler upwelling zone, which has the effect of a weaker transmission of light into the sea. Due to this deficient illumination, phytoplankton come closer to the surface in the coastal zone, while in the oceanic region they remain at a certain depth; this is why the coastal waters appear green and the oceanic waters of the region have an ultramarine colour.

• *The boundary in the vertical plane*. The water layers in the coastal current region between the surface and a depth of 400 m are physically, chemically and biologically different. Indeed, below the surface the waters flowing northwards are of subantarctic origin, while below them there are warm, high salinity waters that move southwards in the form of an intercalated countercurrent that moves between the subantarctic water and the intermediate antarctic water, flows towards the upwelling region, giving up some of its water to compensate for the upwelling water that moves away from the coast in the surface layer.

• *The western boundary of the Peru Current*. The isotherms are oriented from west to east over most of the ocean, but as they approach the coast of South America, they turn northwards until they are parallel to the coast. It is here that the effect of the low temperature of the upwelling waters is felt. This coastal influence is much more pronounced off Peru than off Chile; it is also more evident in northwestern Peru than in the rest of the Peruvian Pacific. The cooling caused by upwelling water is observed as far as the central Pacific region. Off Chile, the effect reaches only 50-130 miles to the west, while off central Peru it is felt up to 150-250 miles. In northwestern Peru, this effect extends farther westward. However, the continuous mixing of the waters causes the western boundary of the Coastal Current to disappear, especially in the winter when the surface waters of the coastal branch of the current meet the waters of the oceanic branch of the current.

• *The southern boundary of the Peru Coastal Current*. ₍Where does the Peru Current begin? As previously indicated, south of 35°S latitude, the subantarctic waters move from west to east (South Pacific drift) until they reach the South American continent. Faced with this natural obstacle that slows its advance, the waters divide: one branch, the Cape Horn Current, heads southwards and the other, the Peru Current, turns northwards, and it is at this point that it originates. The current remains oceanic until it mixes with waters resulting from coastal upwelling. This occurs at latitudes 30°S - 29°S, i.e. near the place where the Subtropical Convergence is closer to the coast and the upwelling water has lower salinity than the adjacent ocean.

• The northern hemisphere of the Peruvian Coastal Current. The cold waters of the Peruvian Coastal Current turn westward in northern Peru and join the South Equatorial Current. Between its waters and the equatorial waters there is a hydrographic boundary. This boundary is recognisable in all transequatorial profiles and represents the true northern boundary of this current towards the equatorial waters of lower salinity and higher temperature. It is in fact a convergence area whose position varies with the seasons.

Dimensions. The width and length of the Peru Current are variable and depend on the flow and transport that takes place in each month and in each year.

The hydrographic structure of the Peruvian Coastal Current. The structure of this current has the following main characteristics.

•	Towards the coast there are relatively cold waters of lower salinity (below 35^ rising to the surface from moderate depths of up to 130 m).

•	On the left side of the current, either on the eastern side or towards the west, there are warmer waters of higher salinity that are likely to sink.

•	At depth there is a current system flowing in the opposite direction to the Peruvian Coastal Current.

Stability. It has its maximum in the winter and its minimum in the summer, being affected by (1) meteorological convection in the winter and (2) high quality water intrusion in the summer. The unstable meteorological conditions in all regions of Peru during the summer impose greater irregularities also on the Peru Current. Thus, in February-March the current becomes weaker and the water temperature rises by about 4°C near the Peruvian coast. The front of the current retreats southwards and the region suffers from the invasion of warmer oceanic waters from neighbouring areas.

Speed. It is wide but relatively shallow, hence slow. It moves its surface waters with a speed of 0.2 to 0.3 knots up to the northwestern region of Peru. Between 5°S and 4°S it leaves the coastal region and increases its speed to about 0.5 - 0.7 knots. It is strongest in August and September in 2 regions, namely in the south between Mollendo and San Juan and in the north between Puerto Eten and Punta Aguja. Both are regions of intense upwelling and in both of them the divergence of the surface waters that tend to move away from the coast under the influence of the Coriolis Force is most evident. The current flows with variable velocity off Callao and somewhat further north. We can state, therefore, that this current increases in velocity towards the north as it receives lateral input from the upwelling and countercurrent. However, the velocity decreases with depth and also in certain areas.

Direction. This current flows with a general equatorward direction along the west coast of South America until about latitude 4°S where it turns northwestward away from the region. Between 15°S and 14°S it also turns offshore, demonstrating that the coastline current influences the general orientation of the current, and where the contour changes its orientation the current also has a tendency to drift away from the coast.

Characteristics. The salinity of the waters of this stream is 34.90% to 35.0%. The temperature varies with the seasons, ranging from 12° to 14°C in the winter and from 15° to 147°C in the summer, and can also reach higher levels. The highest density is found on the right flank of the Peru Current where cooler waters are encountered. On the left side the waters have a lower density. The differences in temperature, salinity and density are maintained due to the vertical and transverse component of the velocity. It is worth mentioning that in the region of the Peru Current there is an absence of dissolved oxygen at moderate depths. The top of the respective layer is near the surface in the region off Callao and deepens to the north and south. The thickness of the respective layer is several hundred metres. The conditions are almost anaerobic.

Significance. It has effects on the climate, the weather in the coastal sea and on the adjacent coast, as well as on the production capacity and distribution of organisms in this part of the South-Eastern Pacific. The observed irregularities in sea conditions, especially in water flow and the occurrence of extreme conditions in certain marine organisms and unexpected effects on the behaviour of marine life and guano birds.

The ***Peruvian Oceanic Current***. The oceanic branch of the Peru Current is located to the west of the coastal branch, well north of 20°S latitude from November to March each year. The flow of this current develops at somewhat higher velocities than that of the Peru Coastal Current.

BIBLIOGRAPHY

POPOVICI, ZACARIAS

Essay on Physical Oceanography. Imarpe, La Punta - Peru , 1966

OCEAN CURRENTS

http://www.inocar.mil.ec/boletin/ecorrientes.html

GLOSSARY

http://www.puc.cl/sw_educ/geo_mar/html/glosario.html#efecto6

WINDS AND OCEAN CURRENTS

http://www.puc.cl/sw_educ/geo_mar/html/h611.html

PACIFIC OCEAN

" http://www.lafacu.com/apuntes/geografia/geograf%C3%ADa_el_oc%C3%A9ano_p

Printed by Books on Demand GmbH, Norderstedt / Germany